MUSEUM OF ANTHROPOLOGY, THE UNIVERSITY OF MICHIGAN

TECHNICAL REPORTS

Number 9

CONTRIBUTIONS IN HUMAN BIOLOGY

No. 2

DIETARY RECONSTRUCTION AT CHALCATZINGO,
A FORMATIVE PERIOD SITE IN MORELOS, MEXICO

by

MARGARET J. SCHOENINGER

ANN ARBOR
1979

c 1979 Regents of The University of Michigan
The Museum of Anthropology
All rights reserved

ISBN 978-0-932206-78-7 (print)

ISBN 978-1-951538-38-5 (ebook)

TABLE OF CONTENTS

LIST OF TABLES.. v

LIST OF ILLUSTRATIONS... vi

PREFACE.. vii

ABSTRACT... x

1. INTRODUCTION... 1

    Dietary Reconstruction.................................... 1
    Food Procurement: Diet, Behavior, Morphology.............. 1
    Archaeology... 5

2. DIETARY RECONSTRUCTION..................................... 9

    Background.. 9
    The Theoretical Basis..................................... 10
        Initial Investigations................................ 10
        Strontium in the Physical Environment................. 10
        Strontium Uptake by Plants............................ 11
        Strontium Metabolism by Animals....................... 12
        Distribution of Strontium within the Animal........... 14
            Portion of the Food Chain......................... 21
        Diagenesis.. 22
    The Technical Basis....................................... 25
    Summary... 29

3. METHODS AND PROCEDURES..................................... 29

    Introduction.. 29
    Strontium Assay in Bone................................... 30
        Sample Preparation.................................... 31
        Analytical Procedures................................. 37
    Analysis of Mortuary Items Associated with Burials........ 41

4. RESULTS.. 41

    Strontium Values.. 49
    Mortuary Analysis... 53

5. DISCUSSION... 59

6. SUMMARY AND CONCLUSIONS.................................... 63

iii

APPENDIX A.   DISTRIBUTION OF STRONTIUM WITHIN ONE BONE......... 63

          Introduction.. 63
          Materials and Methods................................... 65
          Results... 66
          Conclusions... 68

APPENDIX B.   MICROPROBE ANALYSIS................................ 71

APPENDIX C.   ATOMIC ABSORPTION SPECTROMETRY..................... 75

APPENDIX D.   NEUTRON ACTIVATION ANALYSIS........................ 79

          Introduction.. 79
          Irradiation... 80
          Counting.. 81
          An Interference... 82
          Chemical Separation..................................... 83
          Advantages and Disadvantages............................ 85

BIBLIOGRAPHY... 87

LIST OF TABLES

1. Coefficients of Variation of Readings from Microprobe Analysis.....16
2. Bone Sr Levels Compared with Dietary Sr Levels in Rats, Mice and Guinea Pigs...21
3. Accuracy in Trace Analysis 26
4. Results of Analyses for Bone Sr Levels..............................33
5. Correlation Statistics..41
6. Skewness..43
7. Comparison of High Strontium Values Produced by Three Techniques....44
8. Comparison of Diets Through Time....................................49
9. Burials with Accompanying Grave Goods...............................51
10. Coefficients of Variation of Readings from Microprobe Analysis.....68
11. Comparison of Results Produced by Different Data Reduction Methods.85

v

LIST OF ILLUSTRATIONS

1. Contour Map of Chalcatzingo..................................7
2. Cross Section of a Human Radius.............................17
3. Results of Analysis for Ca, P, and Sr Using the Microprobe..18
4. Distribution of Bone Sr Levels in 35 Mink....................20
5. Comparison of Results of Analyses for Bone Strontium Levels..42
6. Levels of Bone Sr in Adults Versus Children..................46
7. Distribution of Bone Strontium Levels in Phase C.............47
8. Distribution of Bone Strontium Levels in Four Temporal Periods at Chalcatzingo..48
9. Cluster Analysis..52
10. Distribution of Bone Strontium Levels in the Plaza Central Versus the Rest of the Site.................................55
11. Area of X-ray Generation Beneath Sample Surface.............64
12. Cross Section of a Human Radius.............................67
13. Results of Analysis for Ca, P, and Sr Using the Microprobe..69
14. Results of Analysis Using Microprobe: Ratios of Elements....70
15. Bragg's Law..73
16. Atomic Absorption Spectrometry.............................76
17. Neutron Activation Analysis................................80

PREFACE

An important theme in Physical Anthropology has been the attempt to unravel the synergistic relation between biological and behavioral adaptations among members of the Order Primates. The results expected from a successful unraveling include both an understanding of each of the adaptations within particular species and also an understanding of how these adaptations functioned during the emergence of our own species. Evidence of this interest can be seen in the recent efforts which physical anthropologists have devoted to the reconstruction of the behavior of past living humans and near-humans.

These reconstructions have often been accomplished using skeletal morphology as an indicator of behavior. For example, posterior tooth size relative to anterior tooth size has been used to separate South African australopithecines into two groups, one mainly herbivorous and the other mainly carnivorous (Robinson 1963). Overall cranial morphology was used by Jolly (1970) to suggest the seed eater hypothesis. Binford (1968) has suggested that hominid evolution between the Mousterian and Upper Paleolithic may be best explained by a shift from an opportunistic hunting system to the more dependable hunting method of following herds of large game animals. Similarly, the morphological changes observed through the Mesolithic may be due to the adaptational requirements of the early hunter/gatherers as opposed to those of the later agricultural groups.

It is obvious that in all of these examples an inferred reconstruction of diet is an interim step between morphology and the behavioral reconstruction. In fact, the ability to reconstruct diet directly would provide important information for explaining certain morphological characteristics and for reconstructing the behavior of human groups. For this reason, this project was organized to investigate a recently suggested means of reconstructing diet. The method, first outlined by Toots and Voorhies (1965), uses levels of bone strontium in order to determine the relative proportions of meat and vegetable matter in the diet. Theoretically, it should provide an important means of adding dietary information in the investigation of problems such as the ones outlined above.

Several people, both within and outside anthropology, have contributed time, equipment, and funds to this project. For providing laboratory space, equipment and advice, I am most deeply indebted to Dr. Dominic Dziewiatkowski and to the members of his staff, especially Marty Brown, Judy LaValley and Myrtle Atkins and to Dr. Jim Christner, all either formerly or presently of the Department of Oral Biology in the Dental School at the University of Michigan. Similarly, I thank Dr. John Eaton and Elaine Berger of the Department of Medicine, University of Minnesota and Dr. Harry Mark of the Department of Chemistry of the University of Cincinnati. The interest of John Jones and Ward Rigot of the Phoenix Laboratories of the University of Michigan and of Thomas Meyers formerly of those laboratories provided a new way of using neutron activation analysis in the analysis of bone strontium. Discussions with Dr. Paul Cloke of the Geology Department greatly facilitated this part of the analysis. Dr. Wilbur Bigelow, John Mardinley, Larry Allard and Peggy Hollingsworth of the microprobe unit and Don Alexander formerly of the Geology Department at the University of Michigan are greatly appreciated for the time they donated during my analysis of bone for strontium using the microprobe.

Discussions with Drs. Jane Buikstra, Carole Bryda Szpunar, and Joseph Lambert of Northwestern University, Robert Gilbert of the University of Southern Mississippi, Noel Boaz of New York University, Gary Wessen of Washington State University and Andrew Fuchs of the University of Pennsylvania have provided useful information concerning the general problems in the trace element analysis of bone. Dr. Hillare Rootare, formerly of the Department of Dental Materials in the Dental School of the University of Michigan and Dr. Ted Lance, formerly of the Department of Chemistry at the University of Michigan helped on separate projects directed toward the synthesis of a strontium hydroxyapatite. Wayne Hruden formerly of Ann Arbor Scientific provided a high purity strontium carbonate for use in analysis. Dr. Richard Aulerich of Michigan State University supplied the sample of mink.

Figures 1, 2 and 12 were drawn by Mark Orsen of the Museum of Zoology of the University of Michigan. Figures 11 and 15 were drawn by Jane Mariouw and the typing was done by Simone H. Taylor, both of whom are members of the Museum of Anthropology of the University of Michigan.

Dr. David Grove of the University of Illinois enabled this particular project to be undertaken by agreeing to let me accompany his crew on their last trip to Chalcatzingo. His subsequent interest and encouragement regarding this project are greatly appreciated. Ann Cyphers and Marcia Merry, both of whom participated in the Chalcatzingo project have generously provided me with unpublished materials from their Master's theses.

Funding for the project came from several sources. An NSF grant (GS-31017) awarded to Dr. David Grove allowed the sample collection. Equipment funds were supplied by the Department of Anthropology, thanks particularly to Drs. Milford Wolpoff and Roy Rappaport. Similarly, chemicals and glassware were purchased from funds supplied by the Museum of Anthropology; for this I thank Drs. Richard Ford and Kent Flannery. Purchase of additional glassware was possible thanks to a Sigma Xi research grant. The neutron activation analysis was performed as part of a University of Michigan staff research grant.

For helpful comments and necessary criticisms special thanks are due Drs. C. Loring Brace and Christopher Peebles of the Museum of Anthropology of the University of Michigan who read earlier drafts of this paper. Also, in this regard, Drs. Kent Flannery, Charles Rulfs and Milford Wolpoff have improved the text with their comments. Finally, I would like to thank Dr. C. Loring Brace for his discussion with me that led to this project and Dr. Antoinette Brown, for her pioneer work in this area and for her continuing encouragement to me.

ABSTRACT

This work was undertaken for the purpose of investigating the use of bone strontium levels in the reconstruction of diet in prehistoric and recent human populations. The skeletal population from Chalcatzingo, a relatively recent prehistoric community in the highlands of Central Mexico, was chosen for study. This sample from a sedentary, agriculturally based community provided a large number of burials from a geographically restricted area. More importantly, three aspects of the community, taken together, indicated that some form of internal ranking was present at Chalcatzingo. These aspects are: 1) the size of the community relative to others in the same valley system, 2) the probable presence of some form of craft specialization and 3) the presence of monumental architecture.

Ethnographic accounts and evidence from the archaeological literature indicate that differential distribution of food stuffs occurs between socially defined ranks. In addition, these accounts report that meat is one of the most highly valued food items. By analogy, then, higher ranking groups at Chalcatzingo probably included more meat in their diets than did lower ranked groups. It was expected that this difference would be reflected in the bone strontium levels. Higher ranked individuals should have lower bone strontium levels than lower ranked individuals.

Relative rank of individuals was determined through an analysis of mortuary items accompanying the burials. Cluster analysis was used to provide a picture of the patterning in the combinations of mortuary items and to depict graphically the groupings of individuals based on the burial items accompanying them.

In the analysis of the bone for strontium the following were considered: 1) the movement of strontium through the physical environment, 2) the uptake of strontium by plants, 3) the metabolism of strontium by terrestrial vertebrates, and 4) the potential for diagenesis. A bone sample from each burial was analyzed by atomic absorption spectrometry. In addition, a subset of the complete set of burials was analyzed by neutron activation analysis. Use of two techniques was necessary because the lack of a nationally recognized standard for bone requires some method of checking on random error. Comparison of results from the two techniques indicates that the values produced by atomic absorption spectrometry reflect the amount of strontium present in the bone. Examination of the pattern of the results and subsequent comparison with the expected variation in bone strontium levels in one species ingesting one diet suggested that more than one dietary regime had been present at Chalcatzingo.

Comparison of the distribution of bone strontium levels with the pattern of social ranking shown by the mortuary analysis demonstrated that the relative level of bone strontium reflected the dietary difference expected based on artifactually defined status differences. The mean bone strontium level for the group of individuals buried with jade beads was lower (more meat) than was the mean bone strontium level for individuals buried without grave goods. A group of individuals buried with shallow dishes had a mean bone strontium level intermediate between the other two groups. It was concluded that the analysis of bone for its trace amounts of strontium is a valid method for reconstructing diet in prehistoric human populations.

1. INTRODUCTION

Dietary Reconstruction

The reconstruction of behavior in prehistoric human populations has often entailed consideration of both the biological and the behavioral adaptations in functional relationship within the process of food procurement. This is an obvious choice for focus because of the basic importance of procuring food for group survival and because of the utility of the known interrelations between its component parts. For all species, those factors which influence the process of procuring food include: (1) the food actually obtained, (2) the behavior (both individual and group) required to obtain the food, and (3) the morphology of the individual (including structural aspects). Because it is the only directly observable component, most behavioral reconstructions are based on initial considerations of morphology. Morphology is the final result of the interaction of three variables: (1) the diet itself, as exemplified by the pathologic case of rachitic bone caused by vitamin D deficiency, (2) behavior, exemplified by robust muscle markings resulting from high levels of muscular activity, and (3) instructions encoded in the genotype.

Morphology has often been used to infer diet, which in turn provides additional evidence with which to reconstruct individual or group behavior patterns. The method of dietary reconstruction using bone strontium levels emphasizes the first variable, i.e., that portion of morphology resulting directly from the diet. It focuses on a minute level of bone morphology, i.e., the microstructure of bone. It will be argued that the amount of strontium included in bone mineral is the direct result of the amount of strontium in the diet. Further, since strontium is distributed differentially between meat and vegetable products the relative amount of bone strontium can be used as an indicator of diet. Once it has been demonstrated to be a feasible method for dietary reconstruction, the information derived from it can help promote the solution of various anthropological problems. The purpose of this project is to provide that demonstration.

Food Procurement: Diet, Behavior, Morphology

Each of the three components (diet, behavior, morphology) contains more than one level, and each may be subdivided in more than one way. Diet can be dealt with generally (e.g., graminivorous or frugivorous) or it can be broken down in several ways. Categories based on calories (high versus normal versus low), meat versus vegetable matter, presence or absence of required amino acids, and the presence or absence of essential vitamins or minerals have all been used for analysis depending on the questions to be answered. Behavior can refer to a group level (called social organization when referring to humans) or to an individual level. Morphology can be used to refer to gross levels of skeletal development: i.e., robust versus gracile, pathologic versus normal, or differences in bone length. Morphology can also be observed on a more detailed level dealing with lines of growth arrest in long bones or on an even finer level, i.e., the actual molecular composition of the bone.

The interactions between components would be expected to relay different kinds of information depending upon the levels that are used. For example, gross skeletal morphology of early hominid jaws and teeth has been used in conjunction with knowledge of resource availability to suggest a general diet and a general form of behavior required in obtaining the diet (see for example Dart 1955; Robinson 1963; Jolly 1970 and Simons 1977, to name only a few). Brace (1977) on a similarly general level has suggested that the behavior involved in the extreme physical requirements of hunting big game was at least part of the reason for the robust skeletal features of Middle and Upper Pleistocene hominids. In contrast, Buikstra (1976 and 1977) and Robbins (1977) have studied a more detailed level of morphology and have used Harris lines of growth arrest to suggest the presence of periods of nutritional inadequacy within certain individuals. At an even finer level of observation, Brown (1973) and Gilbert (1975) have used the molecular composition of bone (specifically bone strontium levels) in order to reconstruct the differences in the meat and vegetable proportions of the diet between individuals within human populations.

In addition to the interaction between diet and morphology, the interaction between diet and behavior has also been the focus of many studies. The method of food procurement practiced by the early hominids of the Plio-Pleistocene probably depended upon some form of population organization in addition to individual behavior patterns. The content and pattern of this group organization have been inferred by analogy with population organization of modern animals that are assumed to be adapting to similar food collection schemes. Both carnivores (Schaller and Lowther 1969; King 1975, 1976) and primates (DeVore and Washburn 1963; Teleki 1974, 1975) have been used as behavioral models.

Social organization (a more complex form of group behavior) within so called "primitive" populations of modern Homo sapiens also appears to be related to the organizational requirements of obtaining food. Binford (following White 1949) has stated that there exists a "generally accepted correlation between forms of subsistence production and societal complexity" (1972:227). Service (1971, also following White) made use of this correlation in outlining his levels of sociocultural integration: band, tribe, chiefdom and state. The band level is considered "the most practical form for foraging peoples" (Service 1971:97, my emphasis). Recently, E. Marx has suggested that the tribe (specifically, the nomadic pastoralists of the Middle East) should "be viewed as a unit

1. INTRODUCTION

of subsistence" (1977:344; my emphasis). The chiefdom level of complexity has been described as one in which certain groups within the community are relieved from direct involvement in food procurement activities. Among such groups, some individuals expend relatively more energy in part time craft specializations while others expend energy in information processing and decision making (see Service 1971; and Peebles and Kus 1977). The state level of social organization comprises specialized groups and the division between the decision making unit and the producers is more discrete than at the chiefdom level (Wright 1977).

Binford's observation concerning the correlation between subsistence production and social organization allows the reconstruction of the general pattern of group behavior once the mode of subsistence is known. This interaction between diet and behavior is on a level equivalent to the interaction between the gross morphological level and generalized group behavior. In order to reconstruct behavior of population subgroups, one must consider the fact that within each human population there is a division of labor and differential distribution of food resources to community members.

Ethnographic accounts have demonstrated that although the differences may be more pronounced on the state level of organization, the division of labor for food procurement is not equitably distributed among members of a population at any of the levels of sociocultural integration. In actuality, different behavior patterns exist for subgroups within any population. In both bands and tribes, age and sex differences determine the divisions of the group for food procurement. Within hunting-gathering bands the women perform most of the tasks involved in gathering plant food while the men do most of the hunting. Young girls and all infants may accompany their mothers, while boys, after achieving a certain skill level, may accompany their fathers. Among tribal communities the mode of subsistence may vary between pastoral and agricultural but the division of labor still follows age and sex lines (Marx 1977). In addition, in both bands and tribes, almost every member of the community is involved in food procurement. As has been previously mentioned, the division of labor for food procurement at the chiefdom or state level includes another variable because certain portions of the population are removed from direct involvement in primary subsistence pursuits.

The link between behavior and diet in human populations is completed with the observation that just as the division of labor for food procurement, production and control falls unequally to members of a population, the distribution of food to members is likewise inequitable. The distribution is mediated through the link of sexual, age determined or social division of labor. In a hunting-gathering population (both band and tribal level), age and sex may govern access to certain food items just as it governs the division of labor. Children may not be fed the same foods as actively working adults. The propinquity of females to plant foods and of males to animal foods may be reflected in a dietary differential between the sexes. For example, among the Anbara aborigines from Arnham Land in Australia, much of the meat from a day's hunt was prepared at "dinnertime camps, in the company of other males" (Meehan 1977a:507); only leftovers were returned to the main camp. Women's gathering groups appear to have had similar "dinnertime camps" (Meehan 1977b).

In settled horticultural or farming communities patterns of resource distribution differ according to the political organization. Within a chiefdom, and more clearly, within a population at a state level of socio-political organization, status and rank differences determine control of and therefore access to certain valued food items. Hatch (1976) summarizes the occurrences of dietary differences in complex chiefdoms of Polynesia and Africa where the "chiefly groups" were able to pick the foods they desired. Early ethnographic accounts from Mexico cited by Spores (1965) support this class prerogative in societies organized on a state level. Zapotec and Mixtec towns in the valley of Oaxaca, at the time of Spanish contact maintained a pattern of dietary differences between their two social classes. One of the most diagnostic items was meat. The consumption of meat "very often reflected class lines" (Spores 1965:967), since the higher status groups controlled its distribution in the same way that non-food high status items were controlled.

The knowledge of the division of labor and the related differential distribution of food-stuffs to group members within human populations should allow a finer reconstruction of social organization of a community once the diet of subgroups within the community is known. Attempts at dietary reconstructions have been undertaken on prehistoric human populations. Since it is the only observable variable in the triad (diet, behavior, morphology) for skeletal populations, morphology has been used in these dietary reconstructions.

In a comparison of artifactually defined status groups with stature, Hatch and Willey (1974) found a positive correlation between status and stature for males within 211 burials from a chiefdom level population. The evidence for dietary differentials within state level organizations is similar; Haviland (1967) discovered that males buried in tombs at Tikal had a higher mean stature than males in non-tomb burial pits. In both studies a genetic basis for the differences was considered unlikely. The authors of both papers concluded that the elite social groups had a higher mean stature because they had a better diet. Haviland suggested that meat was in relatively short supply at Tikal and that access to it was controlled by elites. A dietary differential is probably the best explanation for the stature differential because reduced stature in children has been correlated with protein-poor diets (Garn 1966; Malcolm 1974; among others). Protein-poor diets, ingested throughout the growth period, result in adult stature less than the maximum allowed by the genetic potential. If meat was differentially distributed, as the stature differential suggests, then Tikal is a prehistoric example of the Oaxacan situation where meat was also differentially distributed between social strata.

1. INTRODUCTION

Although suggestive of a dietary differential, the gross morphological level of long bone length is not a fine enough discriminator because of the known genetic component in stature. A finer level of morphology should provide more reliable information concerning the dietary differences between subgroups within a human community. The establishment of dietary differences could then be used to achieve a finer reconstruction of patterned group behavior, that is, the social organization within the community.

The ethnographic references suggest that the dietary categories used when reconstructing dietary differences associated with social rank should be meat versus vegetable since meat appears to be a high status food item. In 1965, Toots and Voorhies presented a method which appeared to reconstruct these dietary categories, although the authors did not suggest its applicability for discriminating between human diets.

They demonstrated the use of micro morphology (bone strontium levels) in discriminating between Pliocene carnivores and herbivores. Brown (1973) and Gilbert (1975) applied the method to human populations for the purpose of discriminating between subgroups ingesting different amounts of meat in an omnivorous diet. Since then Wessen et al. (1977) and Boaz and Hampel (1978) have suggested that the method is not a valid way of reconstructing diet. Theoretical expectations, however, indicate that their conclusions may be somewhat premature. For this reason a thorough investigation was undertaken of the method and its inclusive techniques. As part of this investigation the theoretical expectations are first outlined and discussed. These are based on: (1) the movement of strontium through the physical environment, (2) the uptake of strontium by plants, (3) the metabolism of strontium by terrestrial vertebrates, and (4) the potential for diagenesis. Most importantly the function of strontium in bone growth and development and the position of strontium within bone mineral is discussed. Such considerations allow better formulation of expectations and facilitate interpretation of the results. Secondly, the technical aspects are discussed. An understanding of the analytical techniques is necessary so that any interaction between sample and technique which may produce random error is recognized. Four appendixes discussing such technical problems of measuring bone strontium are included at the end of the text.

Archaeology

The skeletal material chosen for this project is from the population that inhabited Chalcatzingo, Morelos, Mexico during prehistoric times (1150-550 B.C.). This choice was made for several reasons, most important being the fact that a population at Chalcatzingo's level of cultural development should provide a good basis for testing the particular expectations outlined above. Also, and scarcely less important, by chance, the material has been made available to me in considerable quantity. Dr. David Grove of the University of Illinois kindly allowed me to take bone samples from all of the burials recovered from Chalcatzingo. In addition, he directed the arrangements which were necessary to remove any archaeological material from Mexico. Because of David Grove's long-term program of excavation, Chalcatzingo provided a relatively large number of burials from a well defined temporal sequence. Over 90 burials were available for analysis. Most of these individuals probably lived at Chalcatzingo during their entire life span since the community was settled and agriculturally based. Becuase their lives were spent within a restricted geographical area, the non-dietary sources of variation in bone mineral were relatively restricted or at least were generally the same for all individuals. The restricted geographic area also provided a favorable geologic situation. The majority of the burials were recovered from an area which is now and was then drained by one small stream (see Fig. 1) so that all skeletons were treated to the same post-burial effects. The most important reason for choosing Chalcatzingo, however, is because the archaeological material recovered during excavation provided a means of reconstructing the pattern of social ranking.

The archaeological evidence indicates that Chalcatzingo was a settlement of some importance in the Mexican Highlands. It was the largest community in the Amatzinac-Tenango Valley during most of its occupation and covered a minimum of 25 hectares (about 62.5 acres) by the Late Middle Formative period. Monumental public architecture was constructed during the Early Middle Formative period (around 1000 B.C.) and in this same period, the hillside on which Chalcatzingo is situated was terraced (Grove et al. 1976). It is the only site in the Central Plateau known to have Olmec-style bas-relief carvings. Ascertaining the form of the relations between the Olmec of the coast and the inhabitants of the Chalcatzingo in the highlands has provided a basis for much discussion in the literature (Grennes-Ravitz and Coleman 1976; Grove 1968, 1970, 1973, 1974, 1975; Grove et al. 1976; Hirth 1977).

Additional evidence for Chalcatzingo's importance and its levels of social organization is provided by the distribution of structures within the site. During the Middle Formative period a long platform mound was built on the largest terrace. The mound was constructed along the northern edge of the terrace while a series of structures were constructed along its southern edge. One of these structures was larger than house structures on the other farmed terraces, and Grove et al. (1976) think that it represents an elite residence, with the burials beneath its floor representing high status individuals. Another of these structures probably served as a workshop according to its artifact assemblage (Grove et al. 1976; Grove, personal communication). The presence of a workshop suggests that some specialization of labor occurred in the community. A third structure served as an altar. These factors, the relative size of the settlement, the difference in size and function of the structures constructed at the settlement and the presence of labor specialization at least serve to suggest that the community of Chalcatzingo was a chiefdom.

1. INTRODUCTION

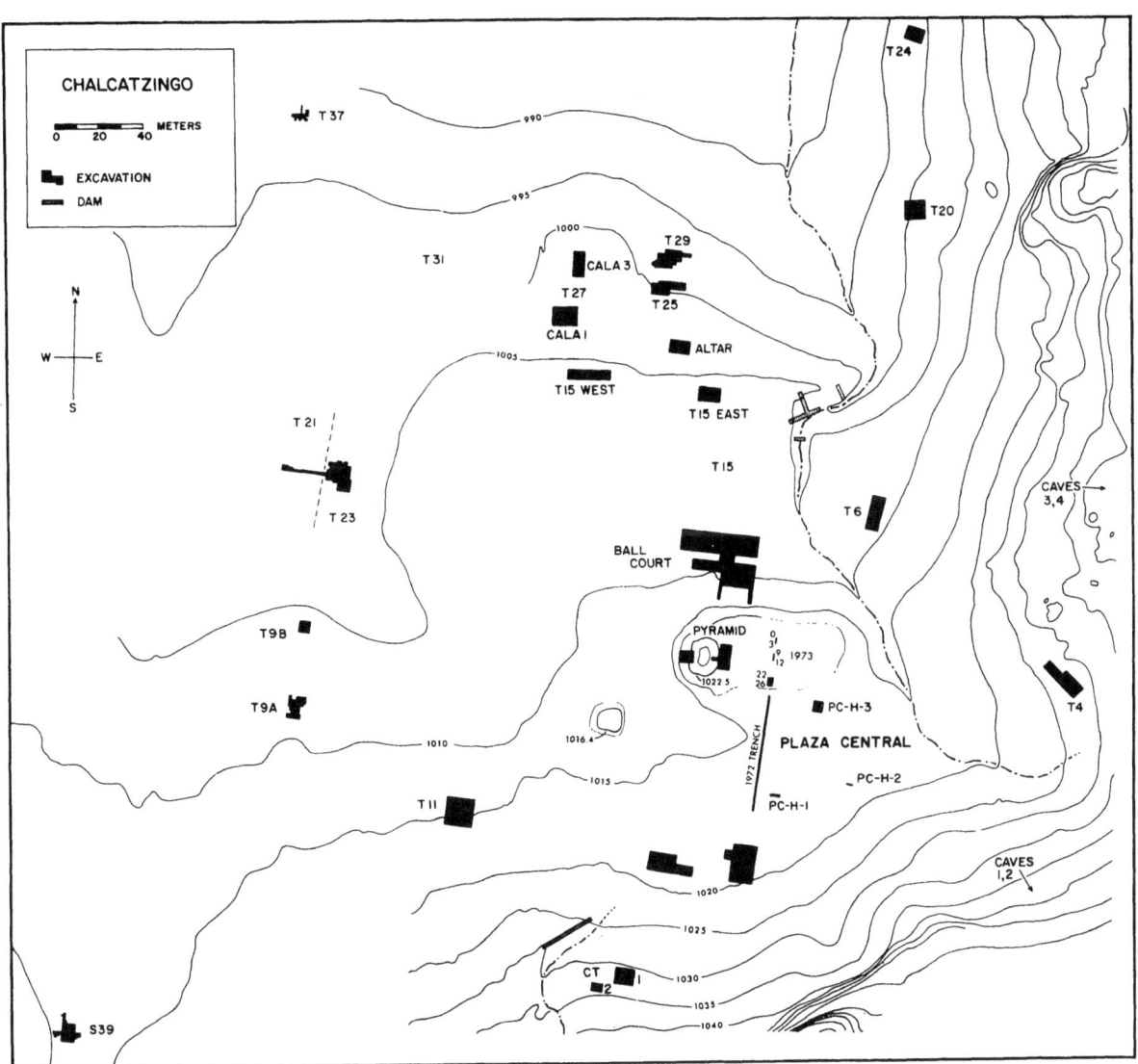

Fig. 1. Contour map of Chalcatzingo. The darkened portions represent the areas that were excavated by David Grove's crew. The small drainage stream is represented by a dashed line to the right of center in the figure. Another small intermittent stream can be seen at the lower left in the figure. A diversion dam directed it to join the other larger stream before its water traversed the site.

The distribution of burials among the different kinds of structures and the distribution of mortuary items accompanying the burials were used to reconstruct the relative social positions of subgroups within the population at Chalcatzingo. A more detailed description of this method is included in Chapter 3.

2. DIETARY RECONSTRUCTION

Background

Human paleontologists and archaeologists usually have attempted to gain dietary information by the collection and analysis of plant and animal remains from archaeological sites and from the catchment areas surrounding such sites. At times, such dietary reconstruction has included very thorough analyses of resources available seasonally for use by a human population (Smith 1975; Marquardt and Watson 1977). There are, however, two major drawbacks to this method. First of all it is limited largely to delineation of "available" rather than "utilized" foodstuffs. Secondly, it reveals information about the population as a whole rather than about subgroups within a population.

Trace element composition of bone has been used to discover aspects of the diet utilized. Toots and Voorhies (1965) demonstrated that strontium levels in bone can indicate the proportion of vegetable material in the diet. These authors were able to place various Pliocene vertebrates in relative trophic position in the food chain based on their skeletal strontium concentrations. Ivanov and Pashkova (1974) have recently confirmed Toots and Voorhies' conclusions in an analysis of modern faunal materials. They suggest the use of trace element levels in bone fragments for species identification. The method was successfully applied to a variety of prehistoric human skeletal populations by Brown (1973, 1974). She hypothesized that individuals in a population lacking social stratification would have a fairly uniform diet while in a highly stratified society there would be more dietary variability. Her contention was supported by the similarity in strontium levels in skeletal material from Middle Woodland period sites in North America and variation in strontium levels in skeletal material from Formative, Classic and Post Classic period sites in Oaxaca, Mexico. Gilbert (1975) and Szpuner (1977) used bone levels of strontium in addition to other trace elements to demonstrate a dietary shift from a broad based resource collecting strategy including both hunting and gathering to a more restricted diet based on maize agriculture in prehistoric American Indian sites. Schoeninger and Peebles (in prep.) have been able to demonstrate a dietary shift from a hunting-gathering population with a large dietary component of fresh water molluscs to a settled agricultural population in a site in northern Alabama.

In opposition, Wessen et al. (1977 and personal communication) believe that the conclusion that strontium is a sensitive indicator of diet has not been adequately demonstrated. In fact, they believe that their own work on herbivores and carnivores suggests the contrary.

Boaz and Hampel (1978) arrive at the same opinion based on their analyses of fossil bone. I think that the reason for these contrary opinions stems from two major sources. The first is derived from the theoretical expectations based on a less than adequate consideration of the movement of strontium through the physical environment and through the trophic levels composed of biological organisms. Wessen et al. (1977) have compared marine carnivores to terrestrial herbivores. Because of the differential in strontium available to marine versus terrestrial vertebrates the two groups are not directly comparable and the food chain argument as presented by Toots and Voorhies (1965) does not apply to this case. The second source of doubt stems from technical problems involved in analyzing trace elements in bone (both modern and fossil). Both of these bases are considered in the remainder of this chapter. The discussion of the analysis of bone strontium will be limited to unfossilized bone because my own analyses were conducted on relatively recent material, although I believe that analytical difficulties may be the reason for the lack of success reported by Boaz and Hampel (1978).

The Theoretical Basis

Initial Investigations

The first major investigations of strontium took place in the 1950s. It was discovered at that time that the radionuclide ^{90}Sr, a fallout product of nuclear weapons testing, was being incorporated into human bones, and that an association between ^{90}Sr and bone tumors could be demonstrated. Various subsequent investigations concluded that the amount of ^{90}Sr in bone was proportional to the amount in the diet (Comar et al. 1955, 1957; Lough et al. 1963; Comar and Wasserman 1964) and that it was virtually impossible to remove the radionuclide from bone (McLean and Urist 1968). These observations suggested that stable strontium, which acts metabolically in the same way as the radionuclide, could be used in dietary reconstructions where dietary differences in strontium intake occurred. Before such a reconstruction can be made, however, the movement of strontium in the physical environment and through the food chain must be understood. The distribution of strontium in the physical environment determines the amount available to the food chain. In general the movement of strontium through the food chain involves the shifts from soil to plants, plants to herbivores and omnivores, herbivores to omnivores, both of these to carnivores, and finally a return to the soil. All of these shifts must receive proper consideration.

Strontium in the Physical Environment

Strontium behaves throughout the sedimentary cycle in much the

2. DIETARY RECONSTRUCTION

same way as calcium because of the similarity in electron configuration, ionization energy and ionic size between these two alkaline earth elements. Calcium and strontium wash down rivers to the oceans where deposits of limestone (calcium carbonate) are formed. In this calcium carbonate, up to 0.24% of the 2+ cation positions are filled by strontium rather than calcium. This small percentage is due, in part, to the difference in relative abundance of the two elements in the earth's crust (calcium is the fifth most abundant element at 3.63 total weight percent; strontium is the 18th most abundant at 0.03 total weight percent [Fleischer 1953]). Under orogenic conditions limestone with its included strontium is uplifted and the cycle begins again as river action begins eroding the newly uplifted sediments (Odum 1971). The strontium cycle appears to be stable since the oceanic strontium/calcium ratio has been of the same order of magnitude at least since the Paleozoic (Odum 1951).

There may be local concentrations of strontium however, since it is also found abundantly with phosphorus in various types of igneous rocks. For this reason strontium and calcium are not necessarily present in the same ratio from one geographical area to the next nor is either element distributed evenly throughout the physical environment.

The actual distribution of strontium should determine the amount available for uptake by plants. Menzel and Heald (1959), however, found that the ratio of strontium to calcium in wheat and alfalfa varied more according to geographical area than according to the type of rock that produced the soil on which the plants were grown. The ratio of strontium to calcium in the plants was very similar to the ratio of strontium to calcium in surface water supplies. The ratio within surface water results from a blending of the ratios within soils "upstream." By either limiting the analysis to populations within one geographical area or by making comparisons between areas only with caution, the variation in bone strontium levels due to the distribution of strontium with the physical environment can be minimized.

Strontium Uptake by Plants

In addition to the variation in the physical environment, the uptake of strontium by particular plants must be considered. In many cases the strontium to calcium ratio in plants is similar to that found in the soil (including surface water) on which they were grown (Bowen and Dymond 1955) and there appears to be little discrimination against strontium as it moves through the plant (Comar et al. 1957). Plants, however, are closed systems, which have no mechanisms for excretion of trace elements. Continued movement of strontium from soil through the plant stem into the leaves during the growing season results in higher concentrations of the element in leaves than in stems. The leaves of trees and shrubs receive elemental nourishment through the growing season and would therefore be expected to have higher strontium levels than grasses, which are more stemlike. Just as expected, Vose and Koontz (1955) found that grasses contain lower amounts of strontium than are contained by leaves. Based on this observation, Toots and Voorhies (1965) were able to discriminate between herbivorous browsers and grazers (identified by tooth morphology) by the higher strontium levels in the former. Storage organs such as roots appear to have roughly the same amount of strontium as leaves, but nuts and grain heads show higher amounts than leaves (Schroeder et al. 1972). The relative differences in the amount of strontium contained by different plant types (grasses versus shrubs) and by different parts of plants (leaves versus nuts) potentially could be used for finer discriminations between animals with contrasting diets.

Strontium Metabolism by Animals

The amount of strontium deposited in the body parts of animals depends on biological factors and on the amount of the element available to the organism. Invertebrates appear to incorporate strontium in a fashion that is different from that of vertebrates. There is evidence that strontium becomes concentrated in the flesh of marine and fresh water molluscs and crustaceans (Odum 1957; Ophel 1963; Schroeder et al. 1972; Kulebakina 1975). Marine vertebrates have higher levels of strontium than terrestrial vertebrates (Rosenthal 1963; Berg 1972). These elevated levels are due to constant contact with higher strontium concentrations in the oceans and because the animals literally breathe as well as eat and drink the ambient level of strontium which provides an additional source of Sr for the blood stream and, therefore, for bone.

Terrestrial vertebrates incorporate strontium in their skeletons in direct proportion to the amount of this element in their diet. Strontium transport across biological membranes, unlike that of calcium is almost exclusively passive so that a constant (though small) percentage passes into the blood stream (this statement is based on in vitro and in vivo studies listed in Comar and Wasserman[1964]). The majority of strontium is excreted renally, with additional small losses occurring due to lactation and placental transfer in pregnant and lactating females.

The evidence from early in vitro studies indicates that "apatite crystals discriminate strongly against strontium when formed under roughly physiological conditions" (Neuman et al. 1963). In vivo studies, however, suggest the opposite. Comar et al. report on the use of ^{45}Ca and ^{85}Sr double-tracer studies "indicating an over-all preferential movement of strontium from blood to bone" (1957:489). More recently, Reeve and Hesp (1976) investigated the appropriateness of using ^{85}Sr as a tracer for bone calcium. They concluded that there is little, if any, difference in uptake between calcium and strontium by the skeleton and that the rate of removal of the two elements is also identical. Consideration of bone as a dynamic substance, helps to resolve these opposing opinions.

Bone is composed of organic and inorganic constituents. The organic portion is mainly protein in the form of collagen. Collagen constitutes 90-96% of the organic matrix while the rest is made up of ground substance (proteoglycans) and a small amount of reticulin plus bound water (McLean and Urist 1968). The amount of organic matter varies between bones of an individual, between the cortical and cancellous portions of a particular bone, and also between the skeletal materials of different species, but in general it averages between 25-35% (Long 1961). The major portion of bone, however, is the inorganic constituent composed of crystals bound to the organic collagen. The crystals are compounds of calcium and phosphate with the structure of an apatite. Apatites normally consist of (1) divalent cations (Ca^{2+}, Sr^{2+}, Mg^{2+}); (2) tetrahedral anionic radicals (PO_4^{3-}); and (3) electronegative anions (OH^-, F^-, Cl^-). In bone these are thought to conform to the formula $Ca_{10}(PO_4)_6(OH)_2$ with substitutions, by the alternatives listed above, being possible in all the relevant positions. Virtually all of the strontium retained in the skeleton is sequestered in this mineral portion.

The inorganic portion of bone does not always fit this neat pattern. The word "apatite" comes from the Greek, "to deceive" (Neuman and Neuman 1969), and the term seems to describe bone mineral very well. It has long been known that a non-crystalline or amorphous calcium phosphate also exists in the inorganic portion of bone, but what proportion it comprises has yet to be determined. Part of this difficulty is due to the transformation of the amorphous fraction to apatite upon removal from the body and also under aqueous preparation conditions required for the electron microscope (Posner 1973). Because of this difficulty much of the knowledge of bone mineral is derived from analogy with synthetic systems and from comparison of a known amount of strontium injected directly into the blood stream with the amount of strontium excreted.

In synthetic systems, the amorphous stage always precedes the crystalline phase (Termine and Posner 1967). The transition between stages takes varying amounts of time depending on the pH of the system (Termine et al. 1970). During this time, ionic exchange can occur between the amorphous crystal and its surrounding fluid. The hydroxyapatite and amorphous molecules have both surface and lattice positions (Neuman et al. 1963). The strontium and calcium in the surface positions readily undergo isoionic and heteroionic exchange. Column positions within the lattice also appear to be susceptible to exchange for a short time, but Neuman et al. (1963) have shown that there is a loss of exchangeability upon crystal aging. They suggest that "since the mobility of an ion in a crystal is roughly proportional to the number of defects and vacancies present" (1963:223), this loss of exchangeability reflects a filling of column vacancies by calcium during maturation of crystals in vitro. Thus, strontium is discriminated against because during the time of exchange it is replaced preferentially by calcium.

The final result of this model, however, does not agree with the results produced by comparison of the amount of strontium retained in vivo. Reeve and Hesp (1976) report that by using double tracers (^{45}Ca and ^{85}Sr) they have determined that the ratio of labeled strontium to calcium excreted in feces and in urine is the same as the ratio of labeled strontium to calcium injected directly into the blood stream.

These observations suggest that uptake and removal of the two elements is virtually identical. Their results are supported by a similar conclusion reached by Marchall et al. (1973) who found that the rates of ionic substitution for strontium and calcium are similar.

Although the crystal chemistry of hydroxyapatite is still a matter of debate, it seems most likely that the early suggestion by Neuman et al. (1963) provides the best explanation for these seemingly divergent opinions. They "speculate that there exists a compartmentalization or membrane separation of the extracellular fluids and the fluids in contact with the mineral crystals...both ions diffuse into the crystal compartment while calcium ion is pumped out at a preferential rate" (1963:224). In this model, the discrimination against strontium by the apatite crystal is offset by preferential removal of calcium resulting in an overall uptake of strontium and calcium proportionate to their concentration in the blood serum.

The loss of exchangeability noted by Neuman et al. (1963) has also been observed in vivo. This decreased potential for exchange is probably due to increased bone mineralization. The space surrounding the initial bone crystals that was originally filled by water is replaced by more hydroxyapatite crystals (McLean and Urist 1968); thus diffusion of ions is restricted. As a consequence the lattice positions within the crystal are no longer available for exchange and, until bone resorption takes place, the system is stable.

The time available for this ionic exchange conceivably might vary between times of bone building either within an individual, between individuals of one species, or between different species. However, this variation in metabolic rates should produce only minor perturbations in the overall retention of strontium. These perturbations, with estimations of their size will be discussed in the next section.

Distribution of Strontium within the Animal Portion of the Food Chain

In this section the movement of strontium between herbivores, omnivores and carnivores will be discussed. Based on the knowledge of the metabolism of strontium by terrestrial vertebrates, outlined in the preceding section, this movement can be expected to be as follows. An herbivore will ingest a relatively large amount of strontium in its diet, most of which will be excreted; the majority of the remainder will move into the skeleton. Strontium is considered a bone seeking element because 99% of the element stored in the body is found in the skeletal system (Schroeder et al. 1972). An omnivore eating meat in addition to plant material will have a lower total strontium intake due to the lower amount of the element in meat. A carnivore will have even less strontium in its diet because most of the contribution by plant material will be lacking. An animal that eats bone will increase its strontium intake but not to the level of an herbivore (Parker 1976). Before strontium can be used as a measure of diet, however, certain potential sources of variation must be considered.

These are first, the variation in strontium content between the bones of one individual; second, the variation within one bone; third, the variation in strontium due to the age of the individual; fourth, the variation in strontium due to individual differences in metabolism; and fifth, the variation in strontium content between two species on the same diet. These will be discussed separately.

The variation within an individual. The studies on humans and other animals which deal with the problem of variation in strontium content between bones of a single individual all agree that there is no difference between calcified tissues beyond that attributed to measurement error. In a report on strontium levels in the parietals, vertebrae, ribs and femora of 38 individuals, Hodges et al. (1950) found no significant difference between bones of the same individual. In another study (Thurber et al. 1958) using human bone samples from all over the world, the strontium content varied by as much as a factor of three between individuals, but there was no difference between bones of a single individual other than the difference expected due to measurement error. Yablonskii (1971, 1973) analyzed bone mineral for several elements in the long bones of human cadavers. Again, the strontium levels did not vary within an individual although they did vary between individuals. Using a different approach Bang and Baud (1972) traced the distribution and incorporation of strontium into bone mineral in vivo. After prolonged administration of a strontium rich diet, strontium was found to be evenly distributed in bone from all areas of the body. Therefore, it appears that bone mineral in any portion of the body reacts in the same manner toward strontium. If strontium is available for incorporation from the blood into the bone, it will undergo heteroionic exchange for surface calcium, isoionic exchange with surface strontium or it will be included in the synthesis of new bone mineral. Different metabolic rates between times of bone deposition throughout the body do not produce perturbations large enough to affect the ratio of dietary strontium to bone strontium.

The variation within one bone. Very little research has been devoted to the mapping of strontium distribution within one bone; therefore much of the evidence provided here is indirect. The amount of bone mineral is lower in trabecular bone than in cortical bone, but Aitken (1976) has demonstrated that the calcium to phosphate ratios are the same in both kinds of bone. This similarity of ratios suggests that the same molecule of bone mineral occurs in both trabecular and cortical bone. Calcium to magnesium ratios are different between trabecular and cortical bone, but this difference is probably related to the role magnesium plays in the calcification process (see Leonard and Scullin 1969). Since strontium, on the other hand, appears to have no functional role in bone, there is no reason to expect differential distribution of strontium. In comparisons of strontium levels among bones of one individual, as mentioned above, there were no differences between either the rib which is mostly trabecular bone, the parietal which also includes a great deal of trabecular bone, or the femur in which the shaft is mostly cortical bone (Hodges et al. 1950). These results also suggest that differential distribution within a single bone is unlikely.

My own work using the electron microprobe to measure strontium, calcium and phosphorus distributions across bone sections indicates that strontium partitioning does not exist. A more thorough discussion of this analysis is found in Appendix A. In brief, cross sections were taken of human rib, radius, and femur. Sampling was done at 300 micron intervals from one periosteal surface to the other. Some paths across the bone included only cortical bone, other paths included both cortical and trabecular bone (see Fig. 2). Figure 3 is a graph of the number of counts (the electrical output from a spectrometer) which, although not identical to concentration, is proportional to it. That is, the relative amount of each element, between areas of the cross section can be ascertained, although absolute amounts can not.

The graph and the coefficients of variation in Table 1 indicate that there is little variation in strontium across any of the sections, irrespective of the kind of bone traversed. These results must be considered as only tentative support for the even distribution of strontium because of the difficulties involved in quantitative microprobe analysis of trace elements in biological materials (Boyde et al. 1961 and Hall 1968; Appendix B of this paper). Additional investigations of this problem would be helpful; however, given the analyses already performed, it does not seem likely that a random choice of bone area for analysis should provide a source of error. In addition, these results indicate that metabolic-rate differences between times of bone building are not large enough to affect the ratio of dietary strontium to bone strontium.

Table 1. Coefficients of variation of readings from areas analyzed by the microprobe. The areas are situated along lines on the surface of a cross section. Each line extends from one periosteal surface to the other.

	Rib			Radius				Femur
Line traversed	1	2	3	4	5	6	7	8
Type* of bone included in each line	C	C+T	C+T	C	C+T	C	C+T	C
Strontium	15	15	11	14	16	12	10	11
Calcium	10	7	21	9	9	10	13	9
Phosphorus	10	8	11	18	19	9	15	29

* C = cortical
 C+T = cortical + trabecular

The variation due to age. The information relating to the variation in strontium due to the age of an individual is inconclusive. Some researchers report that bone strontium levels are higher in children than in adults. Lengeman (1963) reported that human infants between two and nine months of age did not discriminate against strontium in favor of calcium but that by four years of age the discrimination against strontium was the same as in adults. This difference in discriminatory capabilities

2. DIETARY RECONSTRUCTION

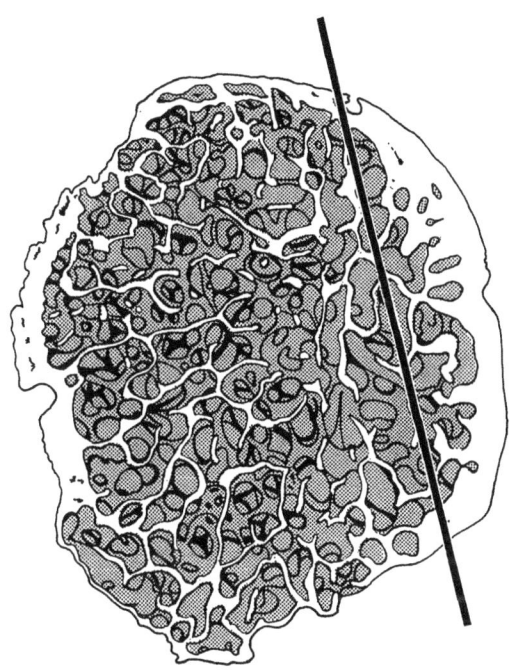

Fig. 2. Cross section of the head of a human radius. This section is typical of the sample used for the microprobe analysis. Readings were taken at 300 micron intervals along a line similar to the one shown on the right side of the figure.

would result in higher strontium levels in young children than in older children and adults. Loutit (1967) found a maximum concentration of strontium versus calcium in human infants one year of age; this ratio decreases with increasing age toward a plateau sometime in adolescence. Brown's data (1973) from a Middle to Late Woddland site (Bussinger) show similar results. The bone strontium levels in adolescents was significantly higher than that in adults. The sample sizes in Brown's study were too small (5 adolescents versus 21 adults) to provide conclusive verification even though the trend was in the expected direction. Lough et al. (1963) agree that strontium to calcium ratios in bone versus diet should be higher for young individuals because of the high absorptive efficiency of sub-adults. Contrary to the preceding results, Turekian and Kulp (1956), Hodges et al. (1950) Alexander and Nusbaum (1959) and Szpunar (1977) found no variation due to age in the strontium to calcium ratio or of the strontium level alone, except in fetal bone. Proposing a third opinion, Bedford et al. (1960) and Sowden and Stitch (1957) found that the overall discrimination against strontium relative to calcium was higher in children than in adults, and that the absorption of strontium from the gastrointestinal tract was 1.6 times higher in adults than in children.

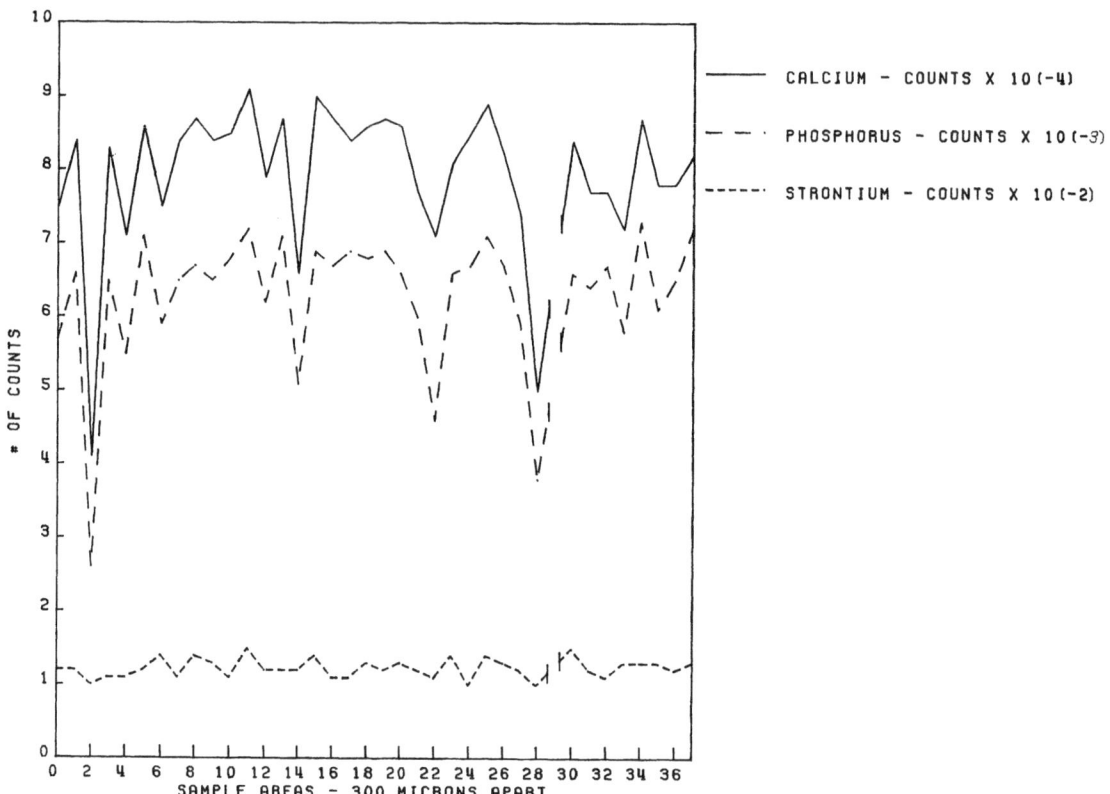

Fig. 3. Digitized electrical output from a spectrophotometer. The output is proportional to concentration; therefore, a change in the number of counts, received from one element, indicates a change in the concentration of that element. The term sampling area refers to the area on which readings were recorded from the spectrophotometer. These areas were spaced 300 microns apart along a line similar to the one shown in Figure 2.

The final answer will have to await further experimental research but the variety of the results reported thus far may in part be explained by the method of data presentation (i.e., as strontium/calcium ratios) in some articles. There is some evidence that the amount of calcium in bone, independent of the amount of bone, decreases with increasing age (Szpunar 1977) and, therefore, is not a constant. In addition, so many elements replace calcium in hydroxyapatite (e.g., Mg, Na, Ba, and K) that a change in calcium concentration will not necessarily be accompanied by an equal and opposite change in strontium concentration. There does not appear, however, to be any change in strontium concentration between maturity and

2. DIETARY RECONSTRUCTION

old age (Szpunar 1977). Data presented by Smith and Smith (1976) suggest that there may be changes in ionic substitutions during aging in human males but this appears (based on x-ray diffraction information) to occur in positions other than that of the 2+ cation. Because of the differences of opinion that exist on the subject, a consideration of the ages of the individuals within the sample is most important.

The variation in metabolism between individuals. The variation in strontium levels between individuals of one species from a single area has been reported to be relatively low; Toots and Voorhies (1965) list coefficients of variation as low as 3.14 ($\bar{X}$=477, S.D.=15, N=4). The highest coefficient they report is 6.46 ($\bar{X}$=526, S.D.=34, N=10). These sample sizes are too small to provide reliable coefficients of variation as indicators of the variation to be expected between members of one species; therefore, it was necessary to analyze another sample. In choosing a sample, known dietary intake was a criterion in addition to large sample size. A sample, meeting both criteria, was made available to me by Dr. Aulerich, the director of the mink farm at Michigan State University. After the mink were killed in the Fall for their skins, Dr. Aulerich saved 35 of them for my use. This group included 22 females and 13 males, all of whom were raised on the same diet. The bones were stripped of the bulk of flesh, boiled, cleaned of any remaining flesh, and finally dried. Since the sample preparation and subsequent analysis was identical for all of the specimens, it was assumed that experimental error had been minimized and that variation between individuals was due to differences in metabolism. Results of this analysis provide a value for comparison with a human sample which would also have individual metabolic differences. Figure 4 presents the strontium levels from the 35 mink humeri. The coefficient of variation is 19.26 ($\bar{X}$=270, S.D.=52). This value, much higher than the 6.46 quoted by Toots and Voorhies (1965), may be a result of the technique that was used (atomic absorption spectrometry) as compared with the technique employed by Toots and Voorhies (x-ray fluorescence spectrometry). It might also reflect greater individual metabolic differences in these small animals (a problem considered below) than in the larger sheep-sized animals analyzed by Toots and Voorhies.

Since the human bone from Chalcatzingo was analyzed by atomic absorption spectrometry, the 19.26 coefficient was the value used as the amount of variation expected within a human population if all individuals ingested the same diet. If there are greater individual metabolic differences in smaller animals, the value of 19.26 can only overestimate the expected variation. Given a choice, the error of using an overestimate is preferable to using an underestimate since the use of the latter might lead to unwarranted subdivisions of a distribution of bone strontium levels with the assumption that these divisions represented discrete dietary differences.

The variation between different species on the same diet. Another problem, alluded to above, is whether metabolic differences between species may affect the strontium deposition independently of diet. Alexander et al. (1956) have provided evidence that this possibility must be considered (see Table 2). In Alexander's project rats and mice (N=6 in both samples) were

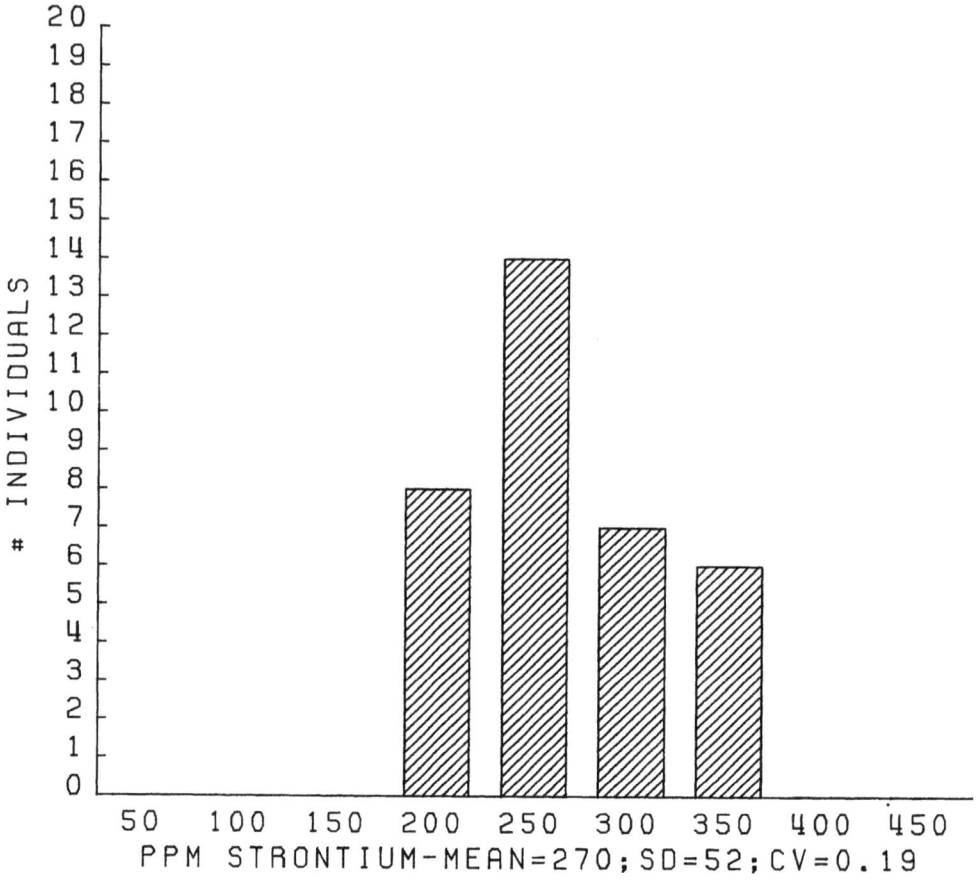

Fig. 4. Parts strontium per million parts bone determined in a sample of 35 mink humeri. All of the mink were fed the same diet during life. The analysis was done by atomic absorption spectrometry.

fed the same diet; guinea pigs were fed a different diet. There is a significant difference (t=3.59, p=0.02) between sample means of bone strontium levels in the rats versus mice. This was true even though their diets were identical. When a comparison was made with the strontium levels in guinea pigs who were fed a different diet, however, this statistically significant difference appears less important. The t-value resulting from comparison of the mice with the guinea pigs is 25.05 and that from comparison of the guinea pigs with the rats is 25.75. In this case the probability that the means are equal is less than 0.01. In addition, the difference is in the expected direction. The guinea pigs were fed a diet high in strontium and had the highest level of this element in their skeletal system. Although the magnitude of a t-value is not an accurate measure of the amount of difference between two samples, this does suggest that although there may be species specific metabolic factors which could produce statistically significant differences in bone strontium where the diet is identical, these differences are of an order of magnitude lower than those produced by dietary differences.

Table 2. Bone Sr levels compared with dietary Sr levels in rats, mice, and guinea pigs (adapted from Alexander et al. 1956).

Animal	Dietary Strontium ppm ash	Bone Strontium ppm ash
Mice	430	190
		240
		220
		170
		260
		250
		$\overline{X}$ 222
		S.D. 35
Rats	430	180
		160
		150
		200
		230
		190
		$\overline{X}$ 185
		S.D. 29
Guinea pigs	2200	1600
		1900
		1700
		1700
		$\overline{X}$ 1725
		S.D. 125

Diagenesis

Finally, there must be a discussion of diagenetic effects -- the postmortem chemical changes in the composition of bone. Strontium is almost completely restricted to the mineral portion of bone (Parker and Toots 1970; Spadaro et al. 1970). For this reason, the processes acting on the organic component (see Hare 1976 for a consideration of these factors) affect the results of strontium analysis only by altering the weight of bone. Such weight alterations can yield inaccurately high or low weight percents of strontium per given weight of bone. If soil replaces the organic matter it should not significantly affect the absolute amount of strontium since the weight percent of strontium in soil is far less than it is in bone (Wyckoff and Doberenz 1968) but it may well lower the amount of strontium relative to bone weight. The difficulty of weight addition is minimized by cleaning the bones prior to analysis which removes most of the adhering dirt. The problem of weight subtraction (removal of organic matter) is minimized by ashing the bone to remove all of the organic matter. This eliminates the uncertainty concerning the amount of organic matter remaining.

The cationic position in bone mineral appears to be generally impervious to diagenesis over a wide range of conditions. Parker (1976) has investigated this factor by comparing the amount of strontium fluorine, and sodium in enamel, dentin, and bone of fossil Subhyracodon. He reasoned that the denser enamel would be less subject to diagenetic changes. Any differences between the enamel, bone, and dentin would indicate a postmortem chemical change since the differences between these tissues in living animals are minimal. The level of fluorine and sodium varied between enamel, dentin and bone. Fluorine content of bone and dentin was higher than in enamel (enrichment) while the level of sodium was lower in bone and dentin than in enamel (depletion). There was no difference in the strontium level in the three types of calcified tissue. Wyckoff and Doberenz (1968) compared strontium content in animal bone from the sites of early humans in the western United States, Pleistocene animals from California and Arizona, Tertiary animals from Arizona and even older fossils from all over the world. Where possible the bone strontium levels in these animals were compared to the bone strontium levels in their modern dietary analogues. There was no significant difference in bone strontium levels between the time periods, but there were significant differences in these levels in animals from different geographical areas and between animals ingesting different diets. In a related project, Nelson (1967) investigated the possibilities of solution and redeposition of strontium in prehistoric clam shells. He concluded that none occurred, indicating that the 2+ cation slot in the carbonate crystal is very stable just as it is in the phosphate crystal of bone. This stability must be due to the bone crystal configuration. The 2+ cation is more tightly bound because it shares two electrons in its bonds whereas both sodium and fluorine share only one. In addition, the strontium atoms occupy positions within the crystal lattice instead of the more easily accessible surface positions. Based on the results of the investigations cited above, there appears to be no reason to anticipate that diagenesis would provide a source of error if bone mineral is retained in the sample.

The Technical Basis

Trace element analysis is a very complex subject. It is impossible to take a sample into "a laboratory" or to analyze it by "a technique" without some knowledge of the sources from which significant error may intrude. One major difficulty in the trace element analysis of bone is the lack of nationally recognized standards for any of the trace elements in bone. The National Bureau of Standards (NBS) has recently certified two biological samples, Bovine Liver (Standard Reference Material 1577) and Orchard Leaves (SRM 1571) both of which took some time to certify. Certification is based on replicability of results by more than one technique. In 1973 Barnard and Dudley reported that the NBS had been able to certify only 6 trace elements in Orchard Leaves. By 1976 Morrison was able to include 14 elements (but not strontium) for the same NBS standard. It had taken nearly three years to expand the list.

Granted that even for experts there are difficulties with trace element analysis, how does this affect the analysis of bone for strontium?

First of all, it requires a decision about the acceptable level of accuracy. For qualitative results (i.e., is the element present?) most of the difficulties are removed, but since all bone (in fact, most naturally occurring calcium compounds) contains strontium, qualitative analysis is not sufficient. I mention it, only because certain techniques are considered "great" for bone but upon further investigation it becomes apparent that they are good only for qualitative answers.

The decision to accept only quantitative results brings up another question: should one accept semi-quantitative or only absolutely quantitative results? For absolutely accurate quantitative results, a nationally recognized standard is a necessity because of the two types of errors inherent in any analysis.

The first type of error is completely random. It can be minimized by standardizing one's sample preparation and analytical procedure but it cannot be eliminated. Expected limits can be placed on it. For example, the Phoenix Lab at the University of Michigan expects a $\pm 3.0\%$ random variation about the "real" value in neutron activation analysis. Although we may never know the "real" value, the next best thing so far is an NBS standard.

The second type of error is systematic. These are errors which may be controlled if their presence is known. They may come from a number of different sources. A major source of error is the result of the interaction between the method of analysis and the sample matrix. Even though a technique may be checked for accuracy by comparing the results it produces on "Bovine Liver" for trace element "X" with the values published by the National Bureau of Standards it can not be assumed to mean that the same technique will necessarily produce accurate results when trace element "X" is in a different matrix. Such an error is not discernible unless comparisons can be made with a recognized standard of the same matrix or with results from another method of analysis (NBS certification method). The only way of evaluating the amount of error is through comparison with a recognized standard since each analytical technique produces its own systematic error.

Another cause of systematic error may be the reagent used (Barnard and Dudley 1973). Even so-called high purity "reagent grade" chemicals have trace impurities. In 1956, Alexander et al. report a reagent grade $CaCO_3$ with 0.097 parts strontium per 1000 parts calcium when the amount of strontium in their samples of bone ash was 0.030 parts strontium per 1000 parts calcium. The same may be true for the acids used in dissolving the sample. Murphy (1976), however, lists the impurities in the acids commonly used for bone dissolution. Strontium is at a level of less than 10 parts per billion of acid; therefore, the acid used probably does not provide much error. One of the major problems caused by impurities in reagent grade chemicals is that of altering the zero or blank level. If the reagent contains a measurable amount of strontium a reading on the blank solution (a standard without any strontium added) will not be a true zero. If the same reagent batch is always used, the error can be removed from the random category and will be systematic so that different analyses by different techniques will be comparable.

A third cause of systematic error can be in the method of sample preparation. Bone samples are usually ashed to remove any interferences from the organic portion. The method of ashing (wet vs. dry) may affect certain of the elements since it may cause volatilization (Gorsuch 1976). Strontium, luckily, is one of the elements listed by Gorsuch as "causing little trouble" (1976:497). For this project, dry ashing was used partially because the sample sizes (0.5-1.0 grams) made wet ashing difficult. In addition, dry ashing is a better method when considering the movement of ions between the sample and the container walls since the atoms of a liquid are more easily exchanged (Speecke et al. 1976). The heat used for ashing was kept low enough (600°C) to prevent the occurrence of reactions between the sample and the container (Coors ceramic crucibles). The quartz tubing used for the sample containers in the neutron activation analysis was high purity Suprasil; it has been analyzed separately and found to contain no strontium. The glass scintillation vials, polypropylene vials, and polystyrene vials used for the dissolution of the bone ash in the analysis by atomic absorption spectrometry have been found to contribute insignificantly to strontium or calcium ionic exchange (Fisher et al. 1976). No dissolved samples were stored longer than four days and during this time they were kept under refrigeration. In the analyses for this project all samples were cleaned, dried and ashed in the same way in order to keep the error introduced through sample preparation at the same level for all samples.

As a result of the two kinds of error inherent in any analysis, I think that accurate quantitative analysis of strontium in bone is impossible without a nationally recognized standard. Henry A. Schroeder attempted to supply reference levels on all the trace elements found in humans. He established a Trace Element Laboratory, and, in a series of articles between 1963 and 1972 in the Journal of Chronic Diseases, he and his associates attempted to analyze the amounts of trace elements in human body parts, in diet and in water. It was quite an undertaking but the difficulty of preparing high purity in-house standards plus the problems in assignment of values for trace elements by consensus within and between laboratories only underlines the need for an NBS bone standard.

Until there is such a standard, those who undertake strontium analysis in bone must minimize the random error by strictly controlling the analytical conditions. If error is kept at the same level on all analyses, comparisons can be made. The most reliable check on random error is the use of at least two methods of analysis (Morrison 1976). The two used in this project were atomic absorption spectrometry and neutron activation analysis. This does not eliminate error since both methods probably add their own systematic error, but if the two methods give the same relative answers they can be assumed to be providing reliable semiquantitative results. Wesch and Bindl (1976) analyzed NBS liver standard 1577 by atomic absorption analysis and neutron activation analysis. The results were compared with the certified values. Although strontium is not included, the results reproduced in Table 3 show the differences

2. DIETARY RECONSTRUCTION

which occurred among sample means and which can be expected between two methods when analyzing liver (the differences may be greater when analyzing bone). Note especially the difference in the parts per million determination of calcium from neutron activation analysis versus atomic absorption spectrometry as compared with the value published by the National Bureau of Standards (NAA = 80 ppm ±30; AAS = 106 ppm ±3.2; NBS = 123). A discussion of the particular problems specific to each technique are discussed in Appendixes C and D.

Summary

Investigation of the theoretical and technical bases underlying bone strontium analysis in dietary reconstruction suggests that such a reconstruction is feasible if certain precautions are taken. First of all, the theoretical considerations indicate that knowledge or control (by choice of sampling area) of the distribution of strontium in the physical environment is necessary to establish a common baseline of the element available to all the plants and animals considered in one's investigation. In addition, knowledge of the differential treatment of the element by plants versus animals, vertebrates versus invertebrates, and marine versus terrestrial organisms allows one to frame more accurate theoretical expectations during the investigation. This differential treatment must be considered in addition to the more often cited expected differences in bone strontium at the herbivore, omnivore and carnivore levels in the food chain. Also, based on literature review and my own supplemental investigations, it would appear that within terrestrial vertebrates certain additional factors should be considered. For instance, limiting the sample to adults is probably wise until the age related differences are more thoroughly investigated. Also, there are some indications of metabolic differences between species that may produce differences in bone strontium levels. It appears, however, that these differences are of an order of magnitude lower than the differences produced between these same species on different diets. Individual metabolic differences have also been considered in order to decide if a given amount of variation actually indicates dietary differences. Within one individual, it is well established that differences in strontium levels between bones are insignificant, and by indirect inference it would appear that differences within one bone are also insignificant. Finally in the last stage, that of postmortem chemical changes, results have been cited which indicate that as long as bone mineral remains, differential removal or addition of strontium does not occur. Since this project deals with relatively recent human skeletal material, I do not believe that there is any question concerning the presence of bone mineral.

The technical basis, i.e., the actual measurement of bone strontium, also provides potential problems. These problems have been discussed in general and the individual problems of technique are discussed in particular in the four appendixes. In addition to the sources of error, two types of error, (i.e., random and systematic) have been outlined. In this project these potential errors were controlled by strict standardization

Table 3. Accuracy in trace analysis (from Wesch and Bindl 1976;234)

Analysis of the NBS liver standard 1577 by means of neutron activation analysis (µg/g dry weight)

Element	Set 1[a] n = 10	Set 2[b] n = 10	Set 3[c] n = 10	NBS[d]
Ca	--	--	80±30	123
Co	--	--	0.21±0.02	0.18
Cu	191±6.2	185±6.8	--	193
Fe	--	--	265±16	275
K	--	--	10400±300	9700
Mg	--	--	659±82	605
Mn	10.2±0.45	10.1±0.5	9.9±0.47	10.3
Mo	3.3±0.3	3.5±0.2	--	3.2
Na	--	--	2330±60	2430
Se	--	--	1.04±0.07	1.1
Zn	124±7.3	127±8.0	123±5	130
Cl	--	--	2550±100	2600

a. Samples ashed prior to irradiation by low temperature ashing.
b. Samples wet ashed after irradiation with H_2SO_4 and H_2O_2.
c. Instrumental neutron activation analysis.
d. Mean values published by the NBS.
a,b,c. Mean ±1 standard deviation.

Analysis of the NBS liver standard 1577 by means of absorption spectroscopy (µg/g dry weight)

Element	Set 1[a] n = 9	Set 2[b] n = 10	NBS[c]
Ca	106±3.2	--	123
Co[d]	0.20±0.016	--	0.18
Cu[d]	186±5.5	188±9.8	193
Fe	272±9.5	266±10	275
K	9600±600	--	9700
Mg	605±32	--	605
Mn[d]	10.3±0.36	9.6±0.6	10.3
Mo[d]	3.4±0.15	--	3.2
Na	2400±200	--	2430
Zn	128±3.6	147±7.3	130

a. Samples low temperature ashed.
b. Samples wet ashed by means of H_2SO_4 and H_2O_2.
c. Mean values published by the NBS.
d. Analyzed by flameless technique. a, b mean.
a, b. Mean±1 standard deviation.

of sample preparation (which minimized random error) and by the use of two techniques which provided a way of checking on random error.

In sum, I believe that with control of the sources for errors in technique and with an understanding of the theoretical basis, bone strontium levels should be able to provide information on certain aspects of the diet. More specifically, it is expected that the level of bone strontium should allow the reconstruction of the relative amounts of meat (or vegetable material) in the diets of individuals within a population. The rest of this paper will be devoted to a test of these expectation.

3. METHODS AND PROCEDURES

Introduction

Two sets of analyses were performed on the Chalcatzingo material. The first set was on the bone itself (trace element analysis discussed previously); the second set was on the burial goods associated with the burials. The second analysis was performed for the purpose of providing a reconstruction of the pattern of social ranking.

All skeletal and cultural material from Chalcatzingo was stored in an old Spanish barracks in Cuernavaca, Morelos, Mexico. During the Spring of 1976, samples were taken from every burial that could be located in the storage area. The bone was very crumbly, indicating that much of the organic matter had been removed. Its poor state of preservation partially explains the fact that of 158 burials reported in field notes only 95 samples were taken. Many of the burials were too fragile for removal during excavation and all archaeological information was taken while the skeletal material was still in the ground. To exemplify the fragmentary nature of the material, in none of the 158 burials was sex determined, and in some cases, even a general age classification of adult versus child could not be decided. The data that were collected were limited mainly to a description of the goods associated with the burials, identification of the time period to which the burials belonged, and a general age classification (adult versus child). The description of burial items was compiled by Merry (1975). The temporal placement of the burials was worked out using ceramic (Cyphers 1975) and radiocarbon analyses. Although some reworking of the time sequence has taken place since (Grove, personal communication) the 1975 chronological divisions described in Cyphers (1975) are used here. Most of the burials are associated with Phase C, Middle Formative period (750-550 B.C.), which is equivalent to parts of the Rosario and preceding Guadalupe phase in the Valley of Oaxaca (Cyphers 1975).

Strontium Assay in Bone

In this project, random and systematic error were monitored by analyzing the samples by two techniques. Because there is some interaction between every analytical technique and the sample matrix (bone in this case), exact replication between the two techniques (i.e., absolute quantitative results) was not expected. That is, for the reasons discussed in Chap. 2 I did not expect the same value of parts strontium per million parts bone to be produced by both techniques (for a discussion of the interaction between technique and matrix see appendixes

B, C and D). Rather, if both techniques ordered the samples in the same way so that a particular sample showed the same relative level of strontium in both techniques, then the results were considered sufficiently accurate. Until a nationally recognized bone sample has been produced, this means of control seems the best way of evaluating accuracy.

Sample Preparation

Each bone sample was placed in a separate glass scintillation vial containing deionized water and was cleaned in an ultra-sound generator for two or three minutes. This is the technique for cleaning bone used by the staff of the Oral Biology Laboratory in the Dental School at the University of Michigan. They have checked for losses of calcium and phosphorus from the bone due to this preparation technique and concluded that there are none (Dr. D.D. Dzwietkowsky, personal communication). It has been assumed, therefore, that strontium will not be affected by cleaning because of its similarity to calcium and also because of Parker's (1976) and Nelson's (1967) reports on the lack of diagenetic effects on strontium. The cleaning via ultrasonic waves was much more thorough than was cleaning by hand. In addition less damage to the bones occurred when the ultra-sound generator was used. It is unlikely that all of the soil was removed from all of the samples but there was probably too little remaining to affect the results greatly. In any event the amount of strontium in the soil at Chalcatzingo was much lower than the amount in bone. An analysis of the soil showed that the strontium level was less than 100 micrograms (10^{-6}g) of strontium per gram of soil. At this level between 0.1 and 0.2 gram of soil would have to adhere to the bone to cause any difficulty given the amount of strontium in the bone (about 700 micrograms of strontium per gram of bone). The total sample was usually about one-half gram (range from 0.1 to 1.0 grams); it would have been unlikely for 20% to 40% of its total weight to be due to soil since all the samples were clean in appearance. The ash/bone ratios were also used as a check on the amount of soil in the bone. Although Long (1961) states that 67% of bone in a human femur is mineral, the ash/bone ratios from Chalcatzingo are much higher (92%). This value is probably due to the postmortem removal of most of the organic matter. I have assumed that in the 2000 years that the bone has been buried, the removal of organic matter has been roughly equal in all samples from the main area of the site. This assumption is based on the time span involved and also on the fact that all of the burials in the main area (which excludes the caves) are in the same drainage system. Given the amount of time involved, ground water removal of organic matter should equally affect all burials. If this assumption is made, the variation in ash/bone ratio should be due mainly to the amount of soil adhering to the sample. The mean ash/bone ratio was 0.92 with a standard deviation of 0.02. Three samples with ash/bone ratios lower than 0.84 were eliminated from further analysis except for the case where the atomic absorption spectrometry (in bone) value was equal to one other value.

This line of reasoning does not mean that there is a complete

3. METHODS AND PROCEDURES

absence of soil in the samples, but the low standard deviation does indicate that, in general, there is an equal addition to all samples. As such it is a systematic rather than a random source of error.

Once the samples were cleaned, they were dried for 24 hours in a 115°C oven. They were weighed twice at 15-minute intervals to assure that they were at a constant weight.

Analytical Procedures

Atomic absorption spectrometry. In the analyses conducted by atomic absorption spectrometry (AAS) two techniques of sample preparation were used. Samples from 51 of the total sample set of 92 burials were prepared according to the first preparation procedure. One sample from each one of the total sample set of 92 burials was prepared according to the second preparation procedure. This was done in order to provide a check on the preparation procedure.

In the first preparation, after the bone was cleaned and dried, 0.5 gram of bone was placed in a glass scintillation vial, and digested, without grinding, in 3.0 milliliters of concentrated hydrochloric acid. Hydrochloric acid was used because it does not depress absorbency of strontium as does nitric acid (Perkin-Elmer manual 1971). The vials were placed in a shaker and left for 24 hours, after which the bone was completely digested. From the dissolved samples 0.2 milliliter of the solution was taken, placed in a polyurethane vial and diluted with 3.8 milliliters of a solution of distilled water which contained 10% trichloroacetic acid (TCA), 1.0% lanthanum and 0.5% potassium. The last two components were added to minimize interference and ionization (Perkin-Elmer manual 1971). The total dilution was 1:60 (1:3 in acid; 1:20 in TCA). The atomic absorption spectrometer was a Varian Tectron. The values for the machine parameters were:

wavelength	460.8 nm
fuel	5.75 (acetylene)
support	5.25 (nitrous oxide)
lamp	10 mA
slit	0.5
vertical position of the burner head	2.5
horizontal position of the burner head	cocked so that the 5µgSr/ml solution read at 50% of the scale

The calibration curve was:

Standards	Scale Reading
10 µgSr/ml	0.920
5 µgSr/ml	0.500
2.5 µgSr/ml	0.250
1.0 µgSr/ml	0.100
0.5 µgSr/ml	0.005

Because the curve was linear only up to 5.0 ug Sr/ml of solution, any sample with a reading higher than 0.500 was diluted and rerun. One milliliter of the 1:60 dilution was added to 3.0 milliliters of the TCA buffer solution. This was a 1:4 dilution or a total dilution of 1:240. The standards were checked several times during the analyses of the unknown samples to make sure that the calibration curve remained stable. The complete set of unknown samples was aspirated three times. The average of three readings on each sample was used in the calculation of parts strontium per million parts bone.

The parts per million value (parts strontium per 10^6 parts bone) was calculated as follows:

$$\text{ppm in bone} = \frac{\frac{\text{scale reading of unknown sample}}{\text{scale reading of 1.0 } \mu g Sr/ml \text{ std.}} \times \text{dilution factor (60 or 240)}}{\text{sample weight}}$$

The results are shown in Table 4.

In the second preparatory method, bone ash instead of bone was analyzed. After the bone was cleaned and dried, a small portion was weighed and ground in an agate mortar. Of this finely ground bone 0.5 gram was weighed into an oven dried Coors crucible and placed in a muffle furnace set at 600°C. The samples were left in the furnace for at least 12 hours. After 12 hours the ash was weighed (after reaching room temparature), then replaced in the furnace for two hours and then reweighed (again after reaching room temperature) to make sure a constant weight was reached. The ash weight was recorded in order to calculate ash/bone ratios. The ash was then placed in glass scintillation vials; two milliliters concentrated HCl were added to the ash and the vials were placed in a warm room on a shaker for two hours after which time most of the samples were completely digested. From each digested sample solution 0.1 milliliter was taken and placed in a polystyrene vial, and to this 4.0 milliliters of a stock solution, consisting of 1.0% lanthanum and 0.5% potassium in deionized water, was added. The total dilution was 1:100 (1:2 in HCl, 1:50 in stock solution). The solutions were then analyzed by a Jarrel-Ash spectrometer with the following machine parameters:

wavelength	460.7 nm
fuel	30.0 psi (acetylene)
support	35.0 psi (nitrous oxide)
lamp	15 mA
vertical and horizontal position of the burner head	set so that the light from the hollow cathode ray tube was centered over the slit in the burner.

3. METHODS AND PROCEDURES

Table 4. Results of analyses for bone Sr levels

Sample	Parts Per Million Strontium in Bone			Ash/Bone
	NAA	AAS(bone)	AAS(ash)	
T25 I	634.72		550.77	.93
II	428.25		447.20	.86
III	591.09		858.46	.93
V	863.35		730.77	.95
VI	812.57		902.56	.89
T25 altar				
1			508.93	.95
2			666.21	.92
3	692.71		573.13	.96
4			910.13	.91
6			688.67	.93
7			747.18	.94
8	644.48		556.25	.89
9			660.00	.94
10			364.53	.92
11			926.55	.91
13			606.52	.90
14			618.04	.93
15			935.63	.88
T37 1	1019.27	1131.42	993.78	.86
2	861.71			.91
3	724.44		752.73	.92
4	749.57		823.16	.92
5	579.51		553.85	.90
6	815.22		765.00	.90
T21 1		1213.85	869.85	.92
N 5 1		911.43	852.63	.90
N 2 1		753.85	575.00	.92
T23 1	tube leaked	623.33	370.42	.96
2	724.73	762.16	652.98	.93
4	927.67*	720.00*	557.66*	.83*
5	411.72	552.86	362.46	.93
Area A				
T9A II	782.35	582.00	537.50	.86
3(=8)	784.65	713.36	718.42	.91
5	520.83	398.18	452.38	.95
6	1035.87	938.18	940.00	.94
9	861.95	833.33	603.75	.92
T9B 3		1000.00	669.77	.92
T4E 1	670.00		606.52	.91
2	589.35		442.13	.93

Table 4 (cont.)

Sample	Parts Per Million Strontium in Bone			Ash/Bone
	NAA	AAS(bone)	AAS(ash)	
Cave 1				
1	1432.66	1363.64	1068.75	.95
2	1321.37	1218.82	1053.25	.94
Cave 4				
1	753.18	827.14	583.53	.96
2	1081.38	954.00	700.00	.95
T24				
3	818.52		606.67	.91
4	742.36		613.61	.94
5	1072.23		950.53	.86
6	677.59		736.00	.92
50#	1129.58		1157.14	.90
T29 1		1080.00	669.77	.90
T11 1*		658.64*	503.45*	.82*
T20				
1	866.88	374.12	748.84	.92
3	918.90	1020.00	816.74	.95
4	733.81	787.50	595.77	.94
4B	755.53	751.76	562.25	.93
5	533.36	693.53	634.88	.91
6	576.15	687.27	463.95	.95
T27				
1			514.77	.96
2			391.88	.95
3			431.82	.95
5			341.33	.96
6			525.95	.94
9			777.29	.91
10			316.67	.95
11			539.32	.86
12			651.00	.93
13			426.81	.95
14			602.88	.95
18			492.86	.92
S39A				
2	1214.22	1111.11	947.06	.92
4	502.49	688.42	461.03	.95
7	838.49	828.57	534.00	.89
50	833.84	708.57	402.86	.94

3. METHODS AND PRECEDURES

Table 4 (cont.)

Sample	Parts Per Million Strontium in Bone			Ash/Bone
	NAA	AAS(bone)	AAS(ash)	
Plaza Central Elite Residence				
72-3			499.43	.92
72-9	568.27	635.29	465.00	.93
72-11	620.85	747.50	529.19	.89
72-12	628.65	590.87	540.00	.92
72-17		235.38	295.38	.96
72-24	545.23	621.11	475.00	.95
72-26	560.15	505.26	533.33	.90
PC pyramid				
73-1	806.94	882.36	600.00	.84
73-2	580.50	650.94	533.89	.93
73-3	653.50	624.26	581.25	.93
73-5	646.22	424.29	448.97	.93
73-6	1064.89	926.84	830.24	.92
73-8	520.22	554.09	524.62	.93
73-9	658.27	621.51	529.81	.95
73-10		563.33	527.78	.95
73-28	788.05	621.05	512.73	.94
73-29	800.92	783.46	568.33	.93
73-30		595.79	424.57	.93
73-35	700.43	665.25	552.19	.93
73-39	780.37	678.82	643.85	.93
73-27	912.19	1127.55	900.00	.90
73-34		561.92	465.00	.93
74-1	793.88*	790.38*	634.48*	.80*
73-19		794.12	606.67	.91
	N=58	N=51	N=91	N=91
	X̄=762	X̄=759	X̄=622	X̄=0.92
	SD=206	SD=232	SD=176	SD=0.02
	CV=27	CV=31	CV=28	CV=2

*Ash/Bone ratio lower than 0.84, therefore value not used in calculation of X̄ or SD.

\# this sample could not be located in the field notes and its burial goods could not be determined; therefore, its value was not used in calculating the X̄ or S.D.

The calibration curve was:

Standard	Scale Reading
7.5 μgSr/ml	0.75
5.0 μgSr/ml	0.50
2.5 μgSr/ml	0.25
0.0 μgSr/ml	0.00

As with the previous method, the standards were checked several times during the analysis and each sample was analyzed three times. The average of the three readings was used in the calculation of parts strontium per million parts bone. No samples gave a reading above 0.75, therefore no further dilutions were necessary. Parts strontium per million parts bone and bone ash were calculated as follows:

$$\text{ppm} = \frac{\frac{\text{scale reading of unknown sample}}{\text{scale reading of 0.1 μgSr/ml}} \times \text{dilution factor (100)}}{\text{sample (ash) weight}}$$

$$\frac{\text{parts Sr}}{10^6 \text{ parts ash}} \times \frac{\text{ash weight}}{\text{bone weight}} \times \frac{\text{parts strontium}}{10^6 \text{ parts bone}}$$

The results are shown in Table 4.

Neutron activation analysis. Fewer samples were analyzed by neutron activation analysis than by atomic absorption analysis because of cost and time. The cost for a University of Michigan affiliate is approximately $2.00 per sample if all of the labor involved in sample preparation is provided by the affiliate. Otherwise, the cost is $40.00 per sample. Time was also an important limiting factor. It took eight months after the submission of the first set of samples before the data reduction method, the interference subtraction described in Appendix D, could be worked out.

All of the samples analyzed by neutron activation analysis (NAA) were in the form of bone ash. This form of sample preparation was used because it had been assumed, a priori, that the most accurate AAS results would be on ash. Although bone or ash should make no difference for NAA it was felt that the sample preparation should be kept as similar as possible between the two analytical methods. Fifty-milligram samples of bone ash were weighed into Suprasil quartz tubes which had previously been boiled for several hours in aqua regia (HCl + HNO$_3$) to insure freedom from contamination. The tubes were sealed over a flame and then etched with the sample number (the numbers of the first set of samples washed off during the immersion in the reactor pool). Twenty-four samples, two strontium standards and a blank were placed in a container and irradiated in the nuclear reactor pool for thirty hours. After cooling for at least three weeks, each sample was counted for two hours. The radio-

3. METHODS AND PROCEDURES 37

isotope used to calculate the strontium concentration was ^{85}Sr (T=62.5 days) which emits gamma rays with an energy of 514 Kev. The intensity of the strontium peak was calculated according to an interference subtraction method, and the parts per million value was calculated by comparison with the standards (see Appendix D for a discussion of this procedure). Just as with atomic absorption spectrometry the parts strontium per million parts of bone were calculated using the ash/bone ratio. Results are shown in Table 4.

Analysis of Mortuary Items Associated with Burials

Binford (1972 following Childe 1951) and others have pointed out a correlation, observed ethnographically, between the complexity of mortuary ritual and the level of social complexity, and have suggested the use of this correlation in the analysis of archaeological data. The correlation occurs because, as Saxe (1970:227) has shown, "in a given disposal domain (societal mortuary ritual) the principles organizing the sets of social personnae (the status and roles of an individual) are congruent with those organizing social relations in the society at large." If follows from this that the mortuary ritual provided for an individual reflects the position of the individual in society during life, because the higher the status of an individual, the greater the number of social relationships which result in a greater corporate involvement at death and a greater investment in mortuary ritual.

Analyses of mortuary ceremony have been conducted for several prehistoric populations (Hatch 1976; Peebles 1971; Saxe 1970). These analyses have usually been conducted for the purpose of determining the particular level of social organization represented by the population. The main purpose in analyzing the mortuary ritual at Chalcatzingo was to discover the presence of groups that had different statuses, relative to each other, during life in this community. The aspects of the mortuary ritual remaining at Chalcatzingo are the artifacts included with the burials; therefore, an analysis of these items was conducted for all burials analyzed for strontium. The distribution of these items should reflect differences in mortuary ritual and thereby reflect the social position of the individual with which the items were buried.

It is difficult, in most cases, to evaluate which burial items were more highly respected or required more energy, and consequently it is difficult to evaluate which burial groups represent high status individuals. For example, did the populace at Chalcatzingo believe one double-looped handle vessel was worth less than, equal to, or more than two shallow dishes (a common burial item)? Perhaps the double-looped handle vessels relate more about the age or sex of the individual (see Bass et al. 1971, for an example of a sex specific burial item) than about the status of the individual within the community. Since this is a skeletal population, an emic approach (i.e., the determination of that which was "regarded as appropriate by the actors themselves." Harris 1968:571) is impossible because the members of the population can no longer verbalize their opinions on the relative worth of artifacts. In cases such as this the anthropologist is limited to the use of etic data, i.e., patterns. A pattern in this case, is a grouping of individuals based on the similarities of their burial goods. This notion is adapted from Sneath and Sokal (1973) where a pattern is defined as "any describable properties of the distribution of operational taxonomic units (OTU's) and groups of OTU's in an artifact-space" (1973:193). The burials are the operational taxonomic units while the artifact space consists of all variables used to describe those burials (e.g., age, sex, presence of burial goods). In the Chalcatzingo sample few data on the age and no data on the sex of the individual burials were available. It was assumed, therefore, that the presence of an object in a burial context, even if it might have been limited to one sex within the population indicated a greater deference toward the individual than toward others who had no grave goods. In addition, all types of grave goods were considered as equal in importance because no evaluation could be made a priori.

The variables used in this part of the analysis were:

Time: Middle Formative
  Phase B, sample size = 8
  Phase C, sample size = 64
Late Formative, sample size = 12
Classic, sample size = 7; Post-Classic N=1; Total=8

Age: Adult
Child

Ceramics: Double-looped handle vessel
Shallow dish
Flaring wall bowl
White ware composite bowl
Olla
Hemispherical bowl
Grey ware composite bowl
Cantarito
Other

Jade: Bead, earspool, other

Groundstone: Mano, metate, molcajete

Figurines

Other

These variables have been selected from a larger set given by Merry (1975). She included other variables, but, in most instances, measurement of these could be made in only a few cases. For example, she recorded the orientation of the head and body, but in many cases could not be sure of the orientation because of the fragmentary nature of the material. Because of the large number of questionable observations, the position variables were not included in this analysis. Merry also listed obsidian

3. METHODS AND PROCEDURES 39

as an item of burial furniture but this item was eliminated from the analysis because there was no information about the type of artifact that was shaped from the obsidian. So many tool types could have been included (i.e., blade, scraper, projectile point, etc) that use of the material would have obfuscated rather than clarified any analysis.

It was decided that use of cluster analysis might reveal patterns among the burials that would not otherwise be immediately apparent due to the number of variables. The choice of a particular technique of cluster analysis was made based on the applicability of the technique to the data set from Chalcatzingo. Whallon et al. (1975) and more recently Christenson and Read (1977) have suggested clustering factor scores rather than the raw data scores so that irrelevant variables that might obscure clusters can be removed. This method, however, requires continuous variables which are normally distributed. It is, therefore, not applicable to the Chalcatzingo data set where the variables are discrete and their distributions are skewed.

Sneath and Sokal (1973) describe various types of polythetic agglomerative clustering techniques. These are Sequential (refers to the presence of a recursive sequence of operations), Agglomerative (begin with n burials and successively group into N less than n sets, ending with one set), Hierarchical (successively fewer sets of burials at each successive stage), Nonoverlapping (burials at any one rank are mutually exclusive) techniques. Within this class of clustering algorithms, Sneath and Sokol (1973:228) recommend use of one of the average linkage techniques (1973:228). These algorithms require computation of a similarity average between an extant cluster and a candidate for inclusion as part of the decision concerned with combination.

The MIDAS statistical package, provided by the Statistical Research Laboratory at the University of Michigan, offers several techniques for calculating this average similarity. A number of these techniques require binary (present/absent) data. Although only a few burial items were present in quantities greater than one, it was felt that the information lost by using only presence/absence might be significant. For this reason the data were retained in their original form as counts within each category; Euclidean Distance was used as the measure of similarity.

In the rest of the decision making function for clustering, MIDAS offers two alternatives for fusion of existing clusters, i.e., minimization of centroid distance or minimization of within-cluster variance. Cormack (1971:332) argues for the latter since it provides a means of minimizing the disorder in the system. This technique chooses the alternative that minimizes within cluster variance (Ward's method) by measuring "the desirability of the particular arrangement of the t (burials) into k<t (clusters) at any one stage" (Sneath and Sokal 1973: 241). Peebles (1974) used Ward's method to organize the mortuary items at the large Mississippean site at Moundville in Alabama and it is discussed more completely in his report.

The clustering of the burials from Chalcatzingo was done by strata for two variables: time period and age at death. The only time period that provided a large enough sample size for clustering was Phase C (Middle Formative); therefore, only burials from this phase were included. In addition, only adults were used in the cluster analysis because of the controversy concerning the metabolism of strontium by subadults.

4. RESULTS

Strontium Values

Although not identical, the distributions produced by the three techniques of strontium analysis are very similar (see Fig. 5). In order to test the similarity between the results produced by the three techniques, several correlation statistics were calculated (see Table 5). The values of the product moment correlation coefficient (r=0.82, 0.81, and 0.84) indicate that high (low) strontium levels produced by one technique tend to associate with high (low) strontium levels produced by either of the other techniques. The "coefficient of determination" (r^2) shows that approximately 64% of the variability in levels produced by one technique can be accounted for by a linear relationship with either of the other two techniques.

Table 5. Correlation statistics

Techniques	N	Corr. Coeff.[a] (r)	Tau-B RCorr[b]	Rho RCorr	Concordance RCorr
NAA vs AAS(bone)	40	0.82	0.59	0.76	--
NAA vs AAS(ash)	57	0.81	0.58	0.75	--
AAS(ash) vs AAS(bone)	51	0.84	0.63	0.80	--
AAS(ash) vs AAS(bone) vs NAA	41	--	--	--	0.85

a. correlation coefficient.
b. Rank correlation

Three coefficients of rank correlation were also calculated because systematic error produced by the interaction between sample matrix and analytical technique could have lowered the similarity in the absolute values produced. The first of these rank correlations, Spearman's Rho, is another product moment correlation coefficient. In this case it is calculated on ranks of the case values produced by two techniques instead of being calculated on absolute case values. The second, Kendall's coefficient of concordance, provides an overall measure of the degree of association existing among the three techniques. According to these two measures of correlation (Rho=0.76, 0.75, and 0.80; Concordance=0.85) the overall pattern of ranks is similar between techniques. It is unlikely, therefore, that the pattern is due to technique artifact, but that it is due to the distribution of bone strontium levels (and the distribution of

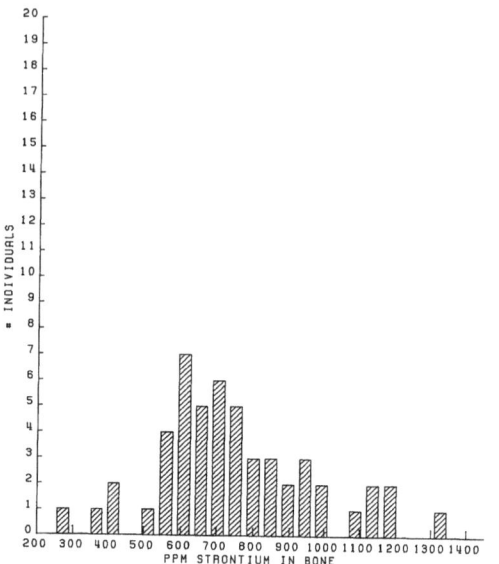

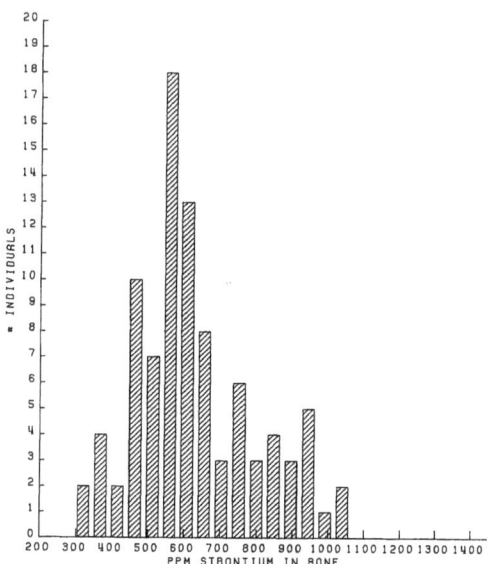

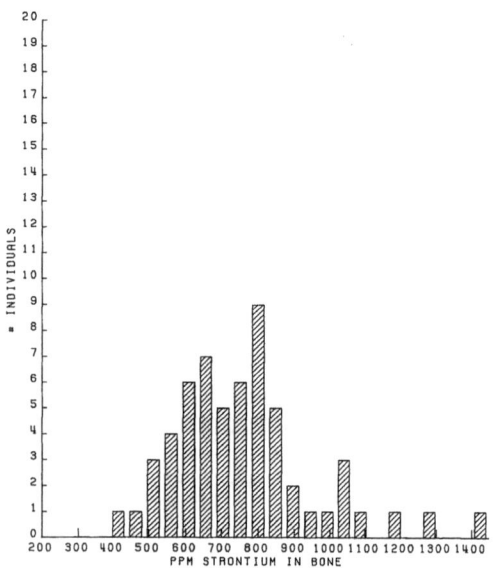

Fig. 5. Comparison of the results produced by three techniques of trace element analysis. The graph at the top left represents the results produced by atomic absorption spectrometry (AAS) on bone that was not ashed (N=51, $\bar{X}$=751, SD=228, CV=30). The graph at the top right represents the results produced by the same technique on bone that was ashed (N=91, $\bar{X}$=622, SD=176, CV=28). The graph to the left represents results produced by neutron activation analysis (NAA) on ashed bone (N=58, $\bar{X}$=762, SD=206, CV=27). The AAS_{bone} and the NAA_{ash} samples are subsets of the complete Chalcatzingo sample. All of the Chalcatzingo samples, however, were analyzed as bone ash by atomic absorption spectrometry.

4. RESULTS

diets) within the population. The third rank correlation Kendall Tau-B is based on the relative ordering of pairs of cases. The level of this coefficient (Tau-B=0.59, 0.58, and 0.63) suggests that ordering of individual pairs of cases is not sufficiently correlated to rely upon the position of one case relative to cases with similar strontium levels.

An interesting point about these distributions is that in each one there is a roughly normal distribution that appears to be skewed toward the right. The right end represents the low meat (high strontium) end of the graph. In order to check this, the three sample distributions were tested for normality by measuring the skewness of each sample. The amount of asymmetry (i.e., skewness) in a normally distributed population is equal to zero while in these three distributions it is: 1.003 in the NAA sample, 0.601 in the AAS on ash sample and 0.495 in the AAS on bone sample (see Table 6). The significance of this deviation from the expected value of the parameter (0.000) was calculated following Sokol and Rohlf (1969:171). The t-values produced and their level of significance are also shown in Table 6. A one-tailed test was used because direction is indicated by the sign of the skewness statistic (in this case it is positive, the direction is toward the right). The probability that the sample analyzed by NAA was drawn from a normally distributed population is less than 0.005, the corresponding probability for the AAS on ash sample is less than 0.01, and the corresponding probability for the AAS on bone is between 0.05 and 0.1. These results indicate that the samples were drawn from a population that is not normally distributed but is instead skewed to the right. The reason for the skewness must be considered.

Table 6. Skewness

Sample	N	Skewness Value	t Value	Level of significance one-tailed test
NAA	58	1.003	3.1971	0.005
AAS (ash)	91	0.601	2.3783	0.01
AAS (bone)	51	0.495	1.4843	0.05-0.1

In general, the same individuals, with a few exceptions, constitute the group with high bone strontium in all three distributions. Table 7 lists the samples that produced high strontium values when analyzed by at least one technique. For AAS on bone ash, all specimens with 800 parts strontium per million parts bone (N=17) were considered to have very high strontium levels. This point was picked because it is approximately midway between the mode in the body of the curve and an incipient mode at the high end of the curve. For AAS on bone, 900 parts strontium per million parts (ppm) bone (N=13) was picked because it would again be at the midpoint of overlap between the main distribution and the incipient mode. The breakpoint chosen for NAA on bone ash was 1000 ppm strontium (N=8) for the same reason given for the two previous distributions. The values produced by the other techniques are also included. A dash indicates that no analysis was performed by the technique for that sample.

Table 7. Comparison of high strontium levels produced by three techniques

Case* No.	Chalcatzingo Number	NAA@	AAS on bone #	AAS on ash ¢
114	S39A #2	1214	1111	947
--	Area A #6	1035	938	940
--	Cave 1 #1	1433	1364	1069
--	Cave 1 #2	1321	1219	1053
138	Cave 4 #2	1081	954	700
104	T24 #5	1072	--	950
107	T37 #1	1019	1131	994
--	T25 III	591	--	858
21	T25 VI	813	--	902
4	T25 #4	--	--	910
--	T25 #15	--	--	936
110	T37 #4	750	--	823
--	N5 #1	--	911	853
95	T20 #3	919	1020	817
--	PC72-27	912	1128	900
87	T9B #3	--	1000	670
34	PC73-6	1065	927	830
11	T25 #11	--	--	927
84	T21 #1	--	1214	870

** All values are in parts strontium per million parts bone.
* Refers to the cluster analysis number; the rest of the individuals were not clustered because they were not Phase C adults.
@ Above 1000 ppm Sr is considered a high strontium level.
# Above 900 ppm Sr is considered a high strontium level.
¢ Above 800 ppm Sr is considered a high strontium level.

Nine samples were analyzed by all three methods, and, of these, six samples are in the high end of all three distributions. The three exceptions are: Cave 4-#2, T20-#3, and PC72-#27. The sample from Cave 4-#2 was one of the few that did not completely dissolve when the HCl was added to the bone ash, and some of the strontium may not have been in solution when the sample was analyzed by AAS on bone ash. Both T20-#3 and PC72-#27 are in the high portion of the AAS on bone and the AAS on bone ash ranges and are above 900 ppm for the NAA distribution therefore they pose little problem since the slightly low NAA values are most likely due to measurement error.

There are seven specimens that were analyzed by two methods only,

4. RESULTS

and, of these, the values for only three agree between the techniques: T24-#5, N5-#1, and T21-#1. Three of the exceptions (T25-III, T25-VI and T37-#4) are the only three samples where the NAA value is less than the AAS on bone ash value. This discrepancy suggests that there may be something wrong with the NAA results for these three samples. The fourth exception is T9B-#3 which is another sample that did not completely dissolve. In summary, of the sixteen samples in the high end of the distributions that were analyzed by more than one technique, nine are ranked high by all the techniques (S39A-#2, Area A-#6, Cave 1-#1, Cave 1-#2, T37-#1, PC73-#6, T24-#5, N5-#1 and T21-#1), two are close enough that the difference may be due to measurement error (T20-#3 and PC72-#27), two contain specimens in which the bone ash did not completely dissolve (Cave 4-#2 and T9B-#3), and three are specimens that have questionable NAA results (T25-III, T25-VI and T37-#4).

The four remaining samples with high strontium levels were run only by AAS on bone ash. The AAS on bone ash values were accepted because in the sixteen samples analyzed by two or more techniques only two (those that were not completely dissolved) had questionable AAS on bone ash results. In the cases where the techniques did not produce equivalent results, the AAS on bone ash value was used when the bone strontium levels were compared with the pattern of social ranking since the results from only one technique should be used in such a comparison.

Because the samples were analyzed by two or three techniques, and because there is general agreement between the techniques, the high values of these nineteen samples are accepted as real rather than as an artifact of a particular analytical technique. It seems reasonable to assume that these individuals had distinctly higher levels of strontium in their skeletal system than did the rest of the population. This conclusion is further supported by the range of variation within the distributions.

All three distributions (NAA, AAS on bone ash, and AAS on bone) have coefficients of variation that are higher than expected if all individuals had had the same diet (0.27-0.30 versus 0.19 from the sample of mink, all of whom had the same diet). In order to determine if the variation is due to the inclusion of infants and children, the bone strontium levels from AAS on bone ash for all samples are compared with those of adults only and with those of children and infants only (Fig. 6). The three distributions are similar, and there is no reduction in the value for the coefficient of variation when children and infants are removed. Because of the lack of agreement about strontium metabolism by preadults, and because there are no specific ages which might make further analysis worthwhile, the infants and children were eliminated from further discussions.

In addition, the high coefficient of variation is not due to the clumping of different time periods. When the strontium levels for Phase C adults (N=47) only are plotted (Fig. 7) the coefficient of variation is still 0.28. When the 10 individuals with the highest bone strontium levels are eliminated from the calculations of mean and standard deviation

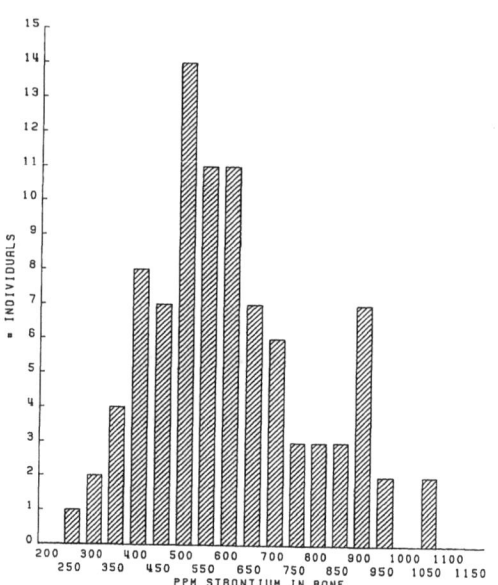

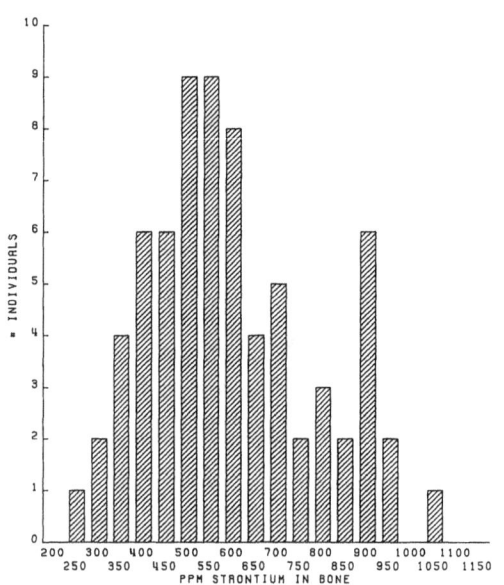

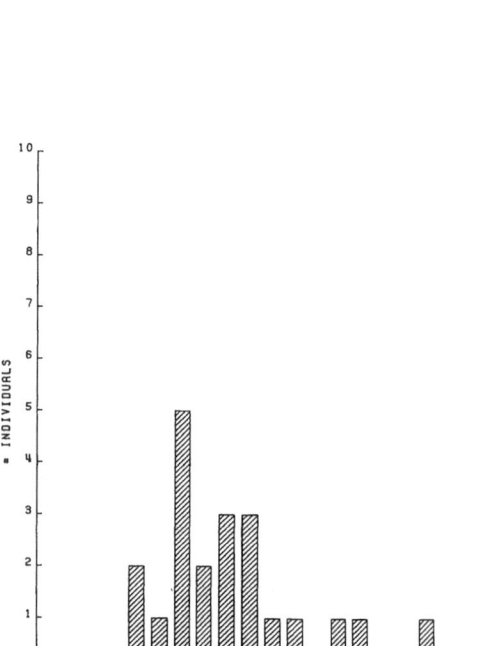

Fig. 6. Comparison of the distribution of bone strontium levels in all of the Chalcatzingo samples with the distribution of adults only and with the distribution of subadults only. The samples were analyzed as bone ash by atomic absorption spectrometry. Graph on the top left contains all samples, N=91, $\bar{X}$=622, SD=176, CV=28. The graph at the top right contains adults only. N=70, $\bar{X}$=622, SD=180, CV=29. The graph to the left contains children and infants only. N=21, $\bar{X}$=633, SD=158, CV=25.

4. RESULTS

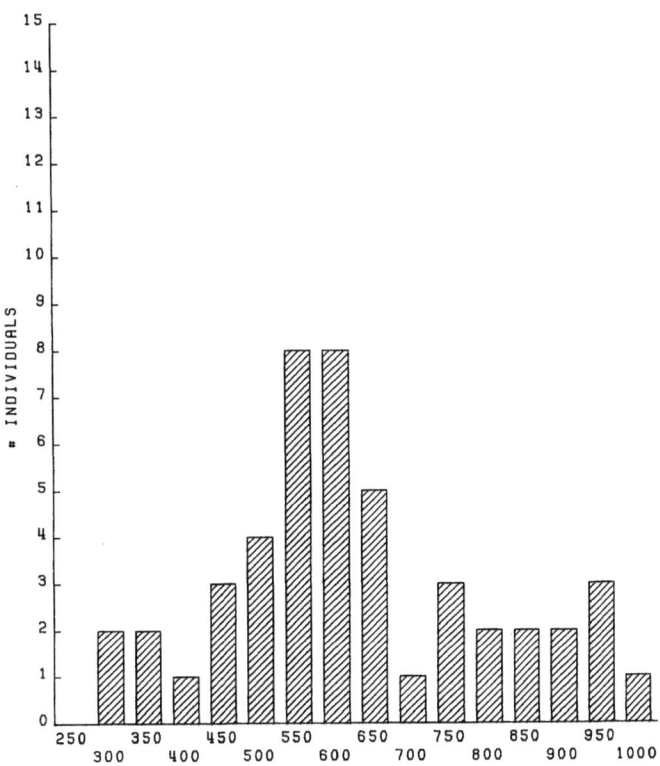

Fig. 7. Distribution of bone strontium levels in adults from Phase C. of the Middle Formative period at Chalcatzingo. The samples were analyzed as bone ash by atomic absorption spectrometry. N=47; $\bar{X}$=627; SD=175.

in the sample of Plase C adults, however, the coefficient of variation drops from 0.28 to 0.20. The latter value is very close to 0.19, which was calculated for the 35 mink, all of whom ingested the same diet.

One other possibility that must be considered is that the variation is merely reflecting dietary shifts during the 200-year span of Phase C. This is possible and the information available on the relative dates of the burials is not good enough to test this. On the other hand, a comparison of the distributions of bone strontium values from Phase B and B-C, Phase C, Late Formative and Classic time periods indicates that on the more gross time scale the dietary distributions remained much the same (see Fig. 8). Admittedly, the sample sizes in the time periods other than Phase C are all too small to provide unquestionable support, but the pattern suggests similarity in dietary distribution. Between any of the distributions, the probability that $\bar{X}_1=\bar{X}_2$ is always greater than 5.0% even in a one-tailed test (Table 8); therefore, the hypothesis of equality of means cannot be rejected. Since dietary shifts were probably not taking place between major time strata, it is suggested that the distribution of strontium values for Phase C was not due to time-related dietary changes.

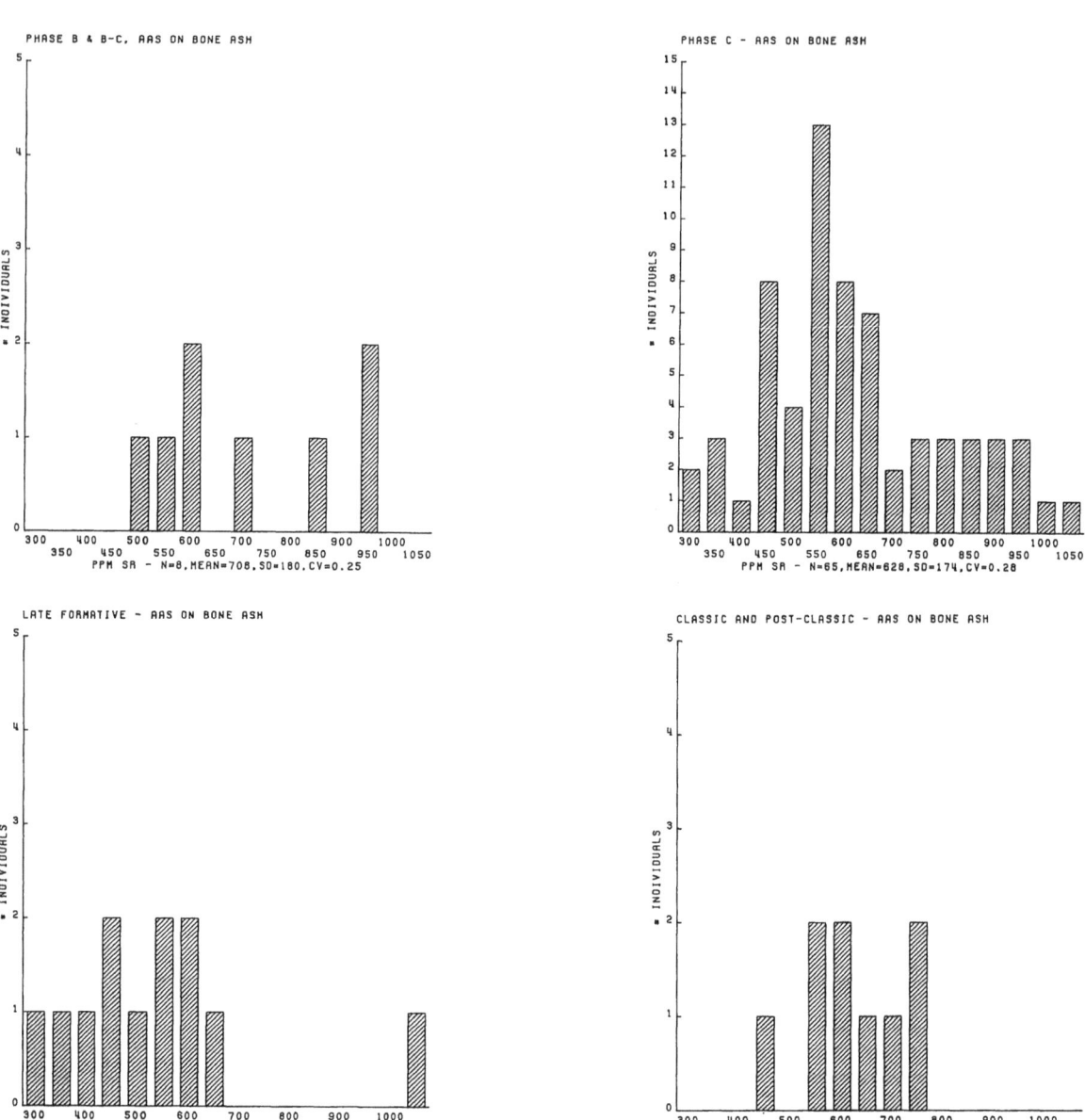

Fig. 8. Comparison of the distributions of bone strontium levels during four temporal phases of the occupation at Chalcatzingo. The samples were analyzed as bone ash by atomic absorption spectrometry. Top left: Phase B+B-C, N=8, $\bar{X}$=708, SD=180, CV=25. Top right: Phase C, N=65, $\bar{X}$=628, SD=174, CV=28. Bottom left: Late Formative, N=11, $\bar{X}$=486, SD=112, CV=23. Bottom right: Classic and Post Classic, N=8, $\bar{X}$=611, SD=106, CV=17.

4. RESULTS

49

Table 8. Comparison of diets through time

Time Periods	T-value	Degrees of Freedom	Significance One-Tailed Test
Phase B & B-C versus Phase C	1.21	71	not significant
Phase C versus Late Formative	1.65	75	not significant
Late Formative versus Classic & Post-Classic	0.88	17	not significant

Mortuary Analysis

Ward's method of cluster analysis produced differing numbers of burial groups depending upon which distance value was chosen (see Figure 9 and Table 9). The greatest difference between distance measurements is between steps 30 and 29. At this point all clusters except one are composed of two or three individuals. The exception is the cluster including cases 4, 11, 20, 21, 27, 36, 49, 55, and 84. The similarity among these individuals is produced by the shared absence of any grave goods. At the next largest difference between distance measurements (between steps 18 and 17) there are three clusters that include three or more individuals. Cluster one, composed of cases 2, 10, 11, 31, and 3 contains individuals accompanied by a jade bead. The individuals in cluster two are the same ones described above with the addition of case #34 accompanied by one mano, case #87 who has one olla and case #103 accompanied by one hemispherical bowl. Cluster three contains cases 47, 57, 107, 67, 110, 77, 95; all of these individuals are accompanied by one or two shallow dishes and little else. The rest of the cases are either in groups of two or have not yet been clustered at all, therefore they will not be discussed. At the next point of a large difference in distance functions, the three clusters described above have been consolidated into one cluster and the remaining clusters are of one or two individuals. In other words all discrimination has been lost.

In considering the "merit" of the clusters, the one with nine individuals buried without any grave goods contains the largest number of individuals joined at the smallest distance. They are, therefore, more similar to each other than are the cases in the three clusters joined at the greater distance. In other words, there is one very good cluster or three fairly good clusters based on minimization of dissimilarity between cases. Three clusters were chosen because it was felt that more information could be provided than if only one cluster was examined. The first cluster has jade beads as its definitive characteristic, the second has no grave goods; the third has shallow dishes.

Table 9. Burials with accompanying grave goods

Case #	Double Looped Handle Vessel	Shallow Dish	Flaring Wall Bowl	White Composite Bowl	Olla	Hemispherical Bowl	Grey Composite Bowl	Cantarito
2	-	-	-	-	-	-	-	-
10	-	-	-	-	-	-	-	-
31	-	-	-	-	-	-	-	-
61	-	-	-	-	-	-	-	1
3	-	-	-	-	2	-	-	-
4	-	-	-	-	-	-	-	-
11	-	-	-	-	-	-	-	-
20	-	-	-	-	-	-	-	-
21	-	-	-	-	-	-	-	-
27	-	-	-	-	-	-	-	-
36	-	-	-	-	-	-	-	-
49	-	-	-	-	-	-	-	-
55	-	-	-	-	-	-	-	-
84	-	-	-	-	-	-	-	-
34	-	-	-	-	-	-	-	-
87	-	-	-	-	1	-	-	-
103	-	-	-	-	-	1	-	-
13	-	1	-	-	1	-	-	-
114	-	1	-	-	-	1	-	-
104	-	-	-	-	-	1	-	-
47	-	2	-	-	-	-	-	-
57	-	2	-	-	-	-	-	-
107	-	2	-	-	-	-	-	-
67	-	1	-	-	-	-	-	-
110	-	1	-	-	-	-	-	-
77	-	1	-	-	-	-	1	-
95	-	1	-	-	-	-	-	1
71	-	1	-	-	-	-	-	1
65	2	-	-	-	-	-	-	-
72	2	-	-	-	-	-	-	-
66	2	1	-	1	-	-	-	1
30	-	1	-	-	-	-	1	1
41	-	-	-	-	-	-	-	1
138	-	-	-	-	-	-	1	-
129	-	2	-	-	-	-	-	-
76	-	5	-	-	-	1	-	1
14	-	-	5	1	-	-	1	-
18	1	-	1	-	-	1	4	-
38	-	-	-	-	-	-	2	-
63	-	-	-	-	1	-	3	-
81	1	-	-	-	-	-	2	-
137	1	-	-	-	-	-	2	1
25	-	1	-	-	-	-	-	1

(Table 9 cont.)

Other Ceramics	Total Jade	Ground Stone	Figures	Other	Strontium PPM(AAS)	Area of Site	Cluster Number
-	1	-	-	-	666	T25	1
-	1	-	-	-	364	T25	1
-	1	-	-	-	581	PC	1
-	1	-	-	-	475	PC	1
-	1	-	-	-	573	T25	1
-	-	-	-	-	910*	T25	2
-	-	-	-	-	927*	T25	2
-	-	-	-	-	731	T25	2
-	-	-	-	-	902*	T25	2
-	-	-	-	-	600	PC	2
-	-	-	-	-	524	PC	2
-	-	-	-	-	529	PC	2
-	-	-	-	-	295	PC	2
-	-	-	-	-	870*	T21	2
-	-	1	-	-	830*	PC	2
-	-	-	-	-	670	T9B	2
-	-	-	-	-	614	T24	2
2	-	-	-	-	606	T25	-
1	-	-	-	-	947*	T39A	-
2	2	-	-	-	950*	T24	-
-	-	-	-	-	465	PC	3
-	-	-	-	-	606	PC	3
-	1	-	-	-	994*	T37	3
-	-	-	-	-	424	PC	3
-	-	-	-	-	323*	T37	3
-	-	-	-	-	320	T23	3
-	-	-	-	-	816*	T20	3
-	2	-	-	-	465	PC	-
-	-	-	-	-	512	PC	-
-	-	-	-	-	552	PC	-
1	2	-	-	-	568	PC	-
1	-	2	-	-	534	PC	-
1	-	1	-	1	449	PC	-
-	-	3	-	-	700	Cave 4	-
-	1	2	-	-	442	T4E	-
1	-	-	-	-	644	PC	-
-	-	-	-	-	618	T25	-
1	-	1	-	-	550	T25	-
1	-	-	-	-	523	PC	-
1	-	1	-	-	533	PC	-
-	-	-	-	-	562	T23	-
-	-	-	-	1	583	Cave 4	-
-	60	-	-	-	634	PC	-

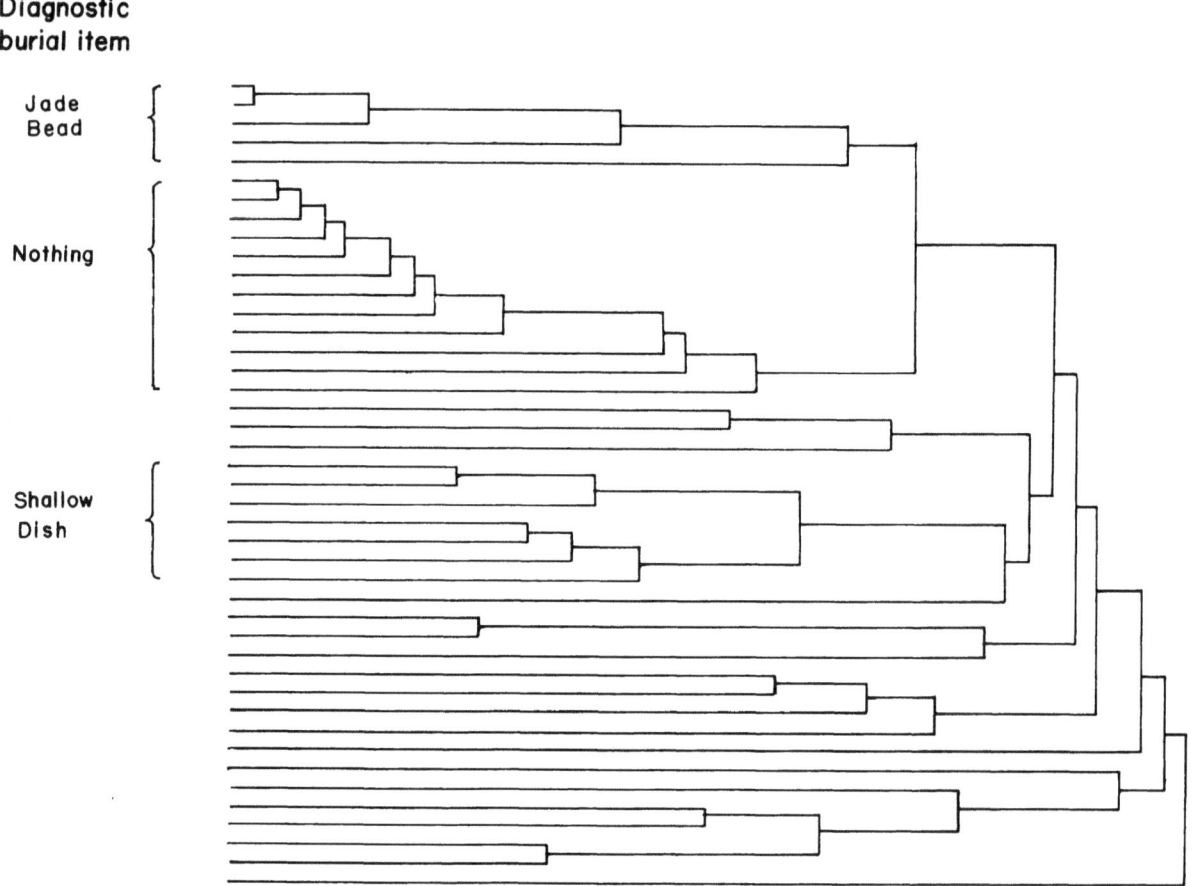

Fig. 9. Results of cluster analysis carried out on the burials at Chalcatzingo. Similarity was calculated using Euclidean Distance. Clusters were combined based upon minimization of within cluster variance. Only adults from Phase C of the Middle Formative period were included in the analysis. Only 43 individuals are included because burial item association was uncertain for 4 of the individuals included in Figure 7. Mortuary items were the variables used for clustering; these included nine types of ceramic vessels, jade, ground stone ware, and figurines. The large difference in distance between step 18 and step 17 prompted the choice of the cluster pattern produced at that position. At this point three clusters were discernible. The burial item diagnostic of each cluster is shown at the left and the burials included in each cluster are enclosed within brackets.

5. DISCUSSION

A comparison of the strontium values with the pattern of social ranking reveals some interesting points of congruence. In general, an attempt to match individual clusters with particular sections of the range of bone strontium values was successful. In the cluster containing individuals accompanied by jade (cluster 1), the mean bone strontium level is 532 ppm (N=5, 364-666 ppm Sr). The cluster defined by the presence of shallow dishes (cluster 3), has a higher mean bone strontium level of 635 ppm (N=7, 320-994 ppm Sr). The cluster containing those cases who had no burial goods and three cases who had one artifact accompanying them (cluster 2) has the highest mean bone strontium level of 700 ppm (N=12, 295-910 pp, Sr). This is in agreement with the status divisions suggested by the mortuary artifacts. Jade is thought by most archaeologists to be a high status item and the bone strontium levels indicate that those individuals (cluster 1) buried with jade at Chalcatzingo probably had a diet containing more meat (shown by lower bone strontium levels) than those individuals in the other clusters. There are some individuals, accompanied by jade plus other grave goods, who were not included in cluster one because of the additional items. When all individuals buried wtih jade are considered, however, the difference in bone strontium levels disappears ($\overline{X}$=610 ppm Sr, N=11, 364-994 ppm). Note however, that the individuals buried with jade but who are not included in cluster one have many burial items in addition to the jade piece(s). Those individuals included in cluster one have no other mortuary items or maybe one other kind of item buried with them and are buried in only two areas of the site, Terrace 25 and the Plaza Central (this will be discussed later). In addition, three out of these five in cluster one were buried with the jade (a bead) in their mouths while none of those outside the cluster had a bead in their mouths. The individuals buried with one jade bead, having few if any other grave items and buried on Terrace 25 or in the Plaza Central probably had more meat in their diets (ppm Sr $\overline{X}$=532; N=5) than those individuals buried with jade, many grave goods and not buried on Terrace 25 (ppm Sr $\overline{X}$=676; N=6).

Of the twelve individuals in cluster two (no grave goods or only one item), six are in the very highest end of the strontium distribution; of these six, five are in the separated high group and one other is in the highest portion of the modal part of the distribution. But there are individuals for whom the high strontium values and the number two cluster do not overlap. Two of these individuals have one grave item accompanying them, which of itself sets them aside from most of the rest of this cluster. There are four individuals included in this cluster, however, who have no burial goods but have low strontium levels.

53

There are also six individuals in the complete sample who have very high bone strontium levels and also have burial goods (one of these is in cluster two). One other variable not included in the cluster analysis may explain this. This variable is the area of the site in which the individuals were buried.

It is unlikely that the variation in strontium level is due to geological differences between parts of the site. As was discussed above, there is evidence from work on both carbonates and phosphates suggesting that strontium is not subject to diagenetic processes. Neither the extremely high nor the lowest strontium values at Chalcatzingo are isolated in particular areas. This distribution argues against post-mortem diagenetic effects producing the pattern of strontium level differences. In addition the high ash/bone ratio ($\bar{X}=0.92$) indicates that bone mineral has not been removed relative to organic matter, and the low standard deviation of the ash/bone ratio (S.D.=0.02) is support for the suggestion that the bone has not been subject to differential treatment in different areas of the site. I think it can be assumed that the area in which the individuals were buried was not contributing directly to the strontium level, but rather that another factor was controlling both the diet and the individual's final resting place.

The four individuals without burial goods who have low strontium levels are from the house structure which Grove called the elite residence (based on its position adjacent to the platform mound and its relatively larger size). Of the six individuals who have burial goods but also have high strontium levels, five come from areas other than the elite residence. The exception is one individual buried with one mano in the Plaza Central. There are no age and sex data which might help to clarify this matter, but even so, one individual does not negate the following general observations:

1) individuals buried outside the Plaza Central without burial goods have high strontium values indicating a diet relatively high in plant material.
2) individuals buried outside the Plaza Central area with burial goods have more variable strontium levels (five out of nineteen have high amounts of strontium while the rest have levels that are in the normal part of the curve) indicating more variable diets.
3) individuals buried in the Plaza Central have strontium levels in the normal part of the Chalcatzingo range no matter what burial goods accompany them (the one exception being the individual buried with only a mano).

When comparing the mean strontium value of those individuals buried in the Plaza Central with those outside that area, there is a significant difference (t=2.74, 42 d.f., the probability that $\bar{X}_1=\bar{X}_2$ is only 0.0005 in a one-tailed test). It is, however, the pattern of the distributions which is most interesting (Fig. 10). The strontium values from outside the Plaza area form a much flatter curve than do the values from inside the Plaza. I think that this can be attributed to a greater dietary variation in those individuals who lived away from the pyramid mound.

5. DISCUSSION

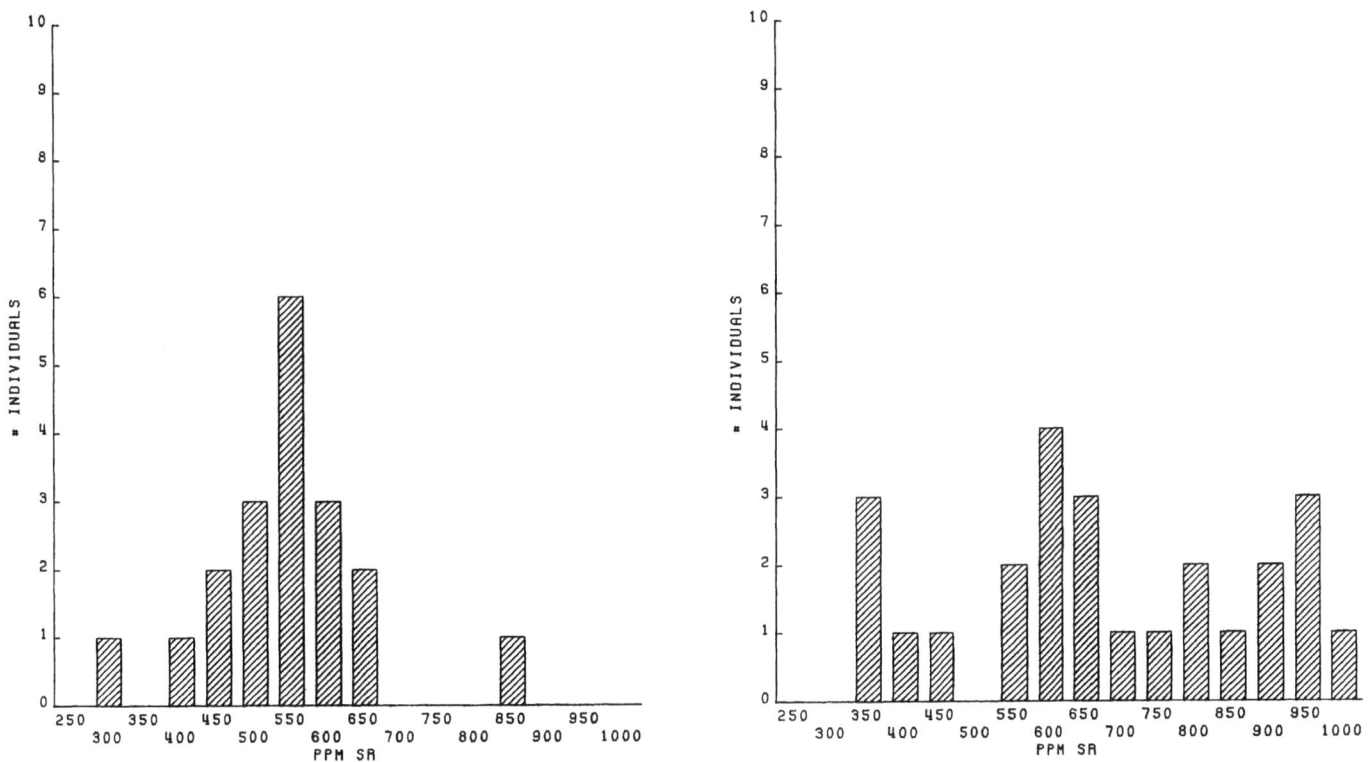

Fig. 10. Comparison of the distribution of bone strontium levels in adults buried in the Plaza Central (left graph) with those buried elsewhere on the site (right graph). Samples were analyzed as bone ash by atomic absorption spectrometry.

The greater variation is produced by the inclusion of a group of individuals whose diets were composed of relatively more vegetable matter (as shown by high strontium levels) than the diets of the rest of the community.

The reason for this variation is not completely clear. Except for two burials interred on the platform mound, the rest of the Plaza Central samples come from subfloor burials interred in the elite residence. If, as Grove has suggested, the persons who lived in this elite residence held the prime authority at the site (Grove et al. 1976), it would follow that they also controlled or had preferential access to valued food items, i.e., animal products. The size of all the residential structures (8 x 10 meters) suggests that they housed extended families (Grove et al. 1976). Some proportion of the residents of the elite house were very likely to have been born into the extended elite family, to have continued residence there, and to have been buried under the floor after death.

There are, however, no discrete groupings in the bone strontium distribution by household such as would be expected if only the residents of the elite house had access to valued food products. One reason for this lack of discrete groups may be post-marital residence. Because individuals probably marry as young adults and because bone turnover rates for adults are low (3.3% to less than 1.0% per year, Bryant and Loutit 1961), the elemental composition of their bone mineral will reflect the diet within their natal residence rather than that of their post-marital residence. Therefore, even if a different dietary regime existed in the post-marital residence their bone would not necessarily reflect the new diet.

More likely, the reason for the lack of discrete dietary groups is that if the individuals in the elite residence controlled certain valued products (both food and other commodities) this control was not absolute. This opinion receives support from the fact that there are burials from other areas of the site which are accompanied by relatively large numbers of items. The situation is more complicated than that of one family in complete control of all valued goods and services to the exclusion of the rest of the community.

Another interesting detail in the association among strontium values, mortuary items and burial location at Chalcatzingo is provided by the distribution of bone strontium levels in the individuals buried on Terrace 25 (see Table 9). Grove et al. (1976) have discussed the presence of an altar on this terrace. It is unlikely that anyone was living in this altar area; rather, it is more likely that individuals from other living areas were brought to the altar to be buried. It would appear that two distinct groups which had different positions in the society of the community were included for burial near this altar. Individuals interred with burial goods have bone strontium levels within the normal range ($\bar{X}=563$; N=6; 364-666 ppm Sr); those without burial goods have high strontium levels ($\bar{X}=868$; N=4; 731-927 ppm Sr). As

5. DISCUSSION

was argued, above, the bone strontium level probably provides more information about the early adult and preadult diet; therefore, it can be inferred that one group buried on Terrace 25 was composed of individuals whose natal family was able to provide them with relatively more meat. These individuals were accorded a particular respect in death as shown by the goods accompanying them and perhaps also shown by the area of the site, an altar area, in which they were buried. The other group consisted of individuals who came from natal families that provided them with relatively less meat, and the individuals were accorded far less status in death based on the burial goods accompanying them.

6. SUMMARY AND CONCLUSIONS

The purpose of the project described in this paper was to demonstrate the feasibility of a method using the level of bone strontium as an indicator of diet. The reconstruction of some aspects of diet from bone micro-structure would provide important information for the attempts by Physical Anthropologists to reconstruct behavioral patterns of prehistoric humans and near humans. Previously, this information has been gathered only inferentially as a result of considerations of morphology or of archaeological data.

The skeletal population from Chalcatzingo, a Formative period agricultural community in the highlands of central Mexico, was used for the study. This sample of burials was chosen partially because it came from a geographically restricted area. Such restriction minimizes variation in strontium levels that would be due to differential distribution of the element in the physical environment. Archaeological information on the site's overall size, distribution of buildings plus the variability of building type, and the probable presence of craft specialization serve to indicate that Chalcatzingo was organized minimally as a chiefdom. Because of this some form of social ranking was expected to have been present at Chalcatzingo. Relative rank of individuals was inferred through a mortuary analysis of the items accompanying each burial. Cluster analysis was used to show the pattern of the distribution of mortuary items among the burials.

Based on both ethnographic and archaeological reports from Africa, the Philippines and Mesoamerica it appears that higher ranked groups of individuals in chiefdoms and states have greater access to high status food items, particularly meat. This should be reflected in bone strontium levels. Specifically individuals of higher rank at Chalcatzingo were expected to have lower bone strontium (higher meat intake) than individuals of lower rank.

Before any actual trace element analysis to test this presumption took place, certain theoretical and technical aspects were considered. First, although there are obviously some differences in the treatment of strontium by various plants and parts of plants, it was concluded that a diet high in vegetable products would provide higher dietary strontium than would a diet containing more meat. Second, as a result of a consideration of the movement of strontium through the animal portion of the food chain, it appears that:

    a. strontium is deposited in bone in proportion to the amount in diet.

b. once bone crystal maturity is attained, removal of strontium occurs only as a result of osteoclastic activity;

c. strontium is distributed evenly between different bones of the skeletal system;

d. strontium is distributed evenly throughout individual bones;

e. there is no consensus of opinion on how metabolic rate differences between adults and children affect strontium deposition. Therefore, the sample in this project was limited to adults;

f. individual metabolic differences may occur. In order to approximate the amount of variation this could produce independent of diet, a sample of thirty-five mink provided by Michigan State University was analyzed for bone strontium levels by atomic absorption spectrometry. All of these animals had been raised on the same diet; therefore, the coefficient of variation from this sample (cv=19.0) was used as the amount expected from one species on one diet;

g. metabolic differences between species produce variations in bone strontium levels on an order of magnitude lower than that produced by dietary differences;

h. the two plus cation position filled by strontium in bone mineral is very stable; diagenesis should not affect bone strontium levels;

The analytical techniques usually used for trace element analysis all display to some extent an interaction with the bone matrix. This interaction produces a certain level of error, which is a problem if the error is random instead of systematic. At present there is no nationally recognized standard of trace element levels in bone which can provide a check on the amount of random error. For this reason two analytical techniques were used and the results compared. The absolute values produced by the two techniques were not identical due to the matrix/technique interaction but the two techniques ranked samples in similar relative positions. Therefore, the rank of the sample was considered to be an indicator of the amount of bone strontium in the sample relative to that in other samples. The coefficient of variation of the Chalcatzingo sample was larger than that expectable if only one dietary regime had been present. In addition, the distribution was skewed toward the high strontium (low meat) end of the graph. This suggests that a group of individuals living in Chalcatzingo were consuming a diet containing less meat than was included in the diet of most community members.

Comparison of the pattern of social ranking, deduced from the mortuary analysis, with the bone strontium levels indicates that this method of dietary reconstruction is feasible. The individuals buried

6. SUMMARY AND CONCLUSIONS

without accompanying mortuary items, assumed to be of low rank, are the same individuals who have the highest bone strontium levels. In fact, these individuals are the ones whose bone strontium levels produce the skewing in the otherwise normal distribution. The group of individuals buried with jade beads, assumed to be an indicator of high rank, has a low mean bone strontium level (higher meat intake). Finally, the group buried with shallow dishes who are assumed to be of rank intermediate between the other two groups has a mean bone strontium level intermediate between the other two groups. The bone strontium levels reflect the dietary difference expected as a result of differential ranking.

The demonstration provided by this project strongly suggests that the analysis of bone for trace amounts of strontium can provide information relating to diet. Using this method, skeletal material can provide a way of performing an independent check on conclusions based on archaeological evidence, and when used in conjunction with archaeological material, analysis for the levels of bone strontium can increase the information derived about prehistoric populations. In addition, such dietary information would be useful in the investigation of several problems of interest to Paleoanthropologists. The australopithecine dietary hypothesis suggested by Robinson (1963) could be tested. The amount of meat actually consumed by Mousterian and Upper Paleolithic hunters might be determined. Finally, tracing the shift in dietary emphasis through the Mesolithic and into the Neolithic would provide information on the introduction of agriculture into Europe.

APPENDIX A

DISTRIBUTION OF STRONTIUM WITHIN ONE BONE

Introduction

Since the skeletal remains from Chalcatzingo were very fragmentary, at times consisting of flecks of bone, limitation of the sample to one type of bone (rib versus long bone) as suggested by Brown (1973) and Gilbert (1975) was rejected as an impossibility. It was equally impossible to limit the sample to only cortical or only cancellous bone because in some burials only the ends of the long bones were recovered, in others only the shafts were preserved, while in others only flecks of both types remained.

It has been determined that the distribution of strontium in the adult skeleton is fairly uniform (Kulp as cited in Thurber et al. 1958). The variation within one bone, however, has never been clearly demonstrated. Because of the lack of choice in bone type at Chalcatzingo, it was deemed necessary to determine whether any area of a bone would be representative of that bone.

This variation within one bone, especially between areas of cortical and trabecular bone, was estimated using electron microprobe analysis. This method was chosen because very small areas, on the order of 10-20 microns could be analyzed. For a discussion of microprobe analysis, see Appendix B. Comparisons from one area to another could then be used to check for any patterned differences which might be present.

The original use of the microprobe in analytical work was limited to materials of homogenous composition i.e. minerals. Recently, however, it has been used for analysis on enamel and dentin (Söremark and Grøn 1966; Frank, Capitant and Goni 1966; and Boyde et al. 1961). It has also been used on bone although less often (Mellors et al. 1966). Even more recently, attempts have been made to use the method for analysis of soft biological tissues (Hall 1968; Russ 1974). These attempts at analysis of biological materials have revealed some problems. One of the first of these, discussed at length by Hall (1968), concerns the lack of comparability between areas of a sample when the areas differ in density. The area of x-ray generation extends beneath the surface of the sample in a roughly teardrop shape (see Fig. 11) and the depth and specific shape of this area varies with the composition of the matrix (including density) and the incident electron velocity. In the analysis of minerals a correction factor can be calculated since the atomic number of the matrix and there-

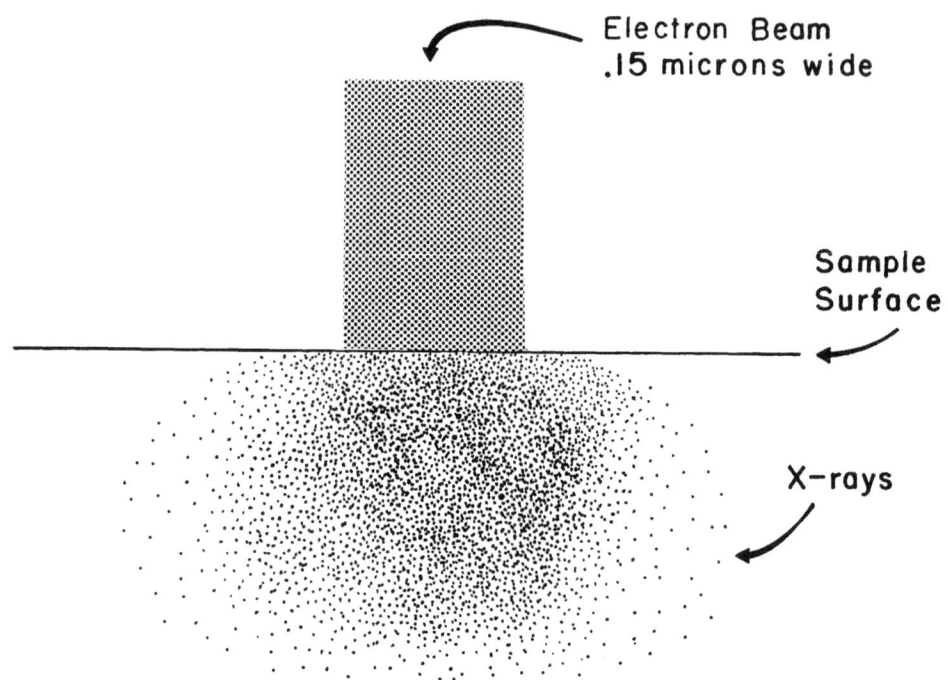

Fig. 11. Representation of the area of x-ray generation beneath the surface of an ideal sample of uniformly low density. The depth and shape of the area are in response to an accelerating voltage of 15Kev electrons.

fore its density can be considered to closely approximate a constant between areas of the sample. This regularity is not necessarily true in biological materials, however, where the density may vary greatly between areas of the sample, especially in soft tissues. Although bone is not a soft tissue, it was decided that an initial assumption of constant density could not be made.

A means of compensating for density differences was worked out using the known relationship between the intensity of the bremsstrahlung (background radiation) and the local total mass per unit area (density). Because the two are proportional to one another (Hall 1968; and Russ 1974), the background counts were held constant during sampling. This has the same effect as holding the density constant; therefore, the area of x-ray production should be comparable between areas of the sample. The Hall technique of using ultrathin samples (5-10 microns thick) was not used because of the difficulties inherent in preparing non-decalcified bone in this manner.

The depth to which the base of the teardrop extends beneath the surface of the sample depends upon the accelerating voltage of the electron

APPENDIX A

beam. The higher the voltage the deeper the penetration beneath the surface, which results in x-rays being produced deeper in the sample. When passing through matter, x-rays are absorbed exponentially with distance (Russ 1974), therefore, the deeper the x-ray is within the sample, the less chance there is of it reaching the surface and entering the detector. A voltage of 15 kili-electron volts was used since it produced a large enough number of counts to be statistically reliable and yet was low enough to avoid interference from absorption.

Another problem was the detection limits of the electron microprobe. For the heavier atomic weight elements, in a relatively light molecular weight matrix a 0.01% concentration (100 ppm in weight ratio) is considered the lower limit of detectability. Since the amount of strontium expected was somewhere between 0.01 and 0.2% (100-2000 ppm, weight ratio) a preliminary test was made. A celestite crystal (0.07% strontium) was used to establish the exact energy at which strontium's characteristic x-rays were detected. The diffracting crystal was then set in the correct position for diffracting the x-ray photons to the detector. The manual from the National Bureau of Standards was checked to see if any other elements emitted x-rays at the characteristic strontium wave length. No element contained in bone emits any photons which could be detected and confused with strontium counts. The counts can be assumed to be due to the strontium concentration of the bone.

Materials and Methods

One cross section of a human rib, two cross sections of a human radius and one cross section of a human femur were made using a mineral saw. The samples were embedded in resin. Before the resin had completely hardened, the samples were placed in a bell jar equipped with a vacuum pump. This step insured permeation of the sample by resin. After the resin had hardened completely (about 12 hours) the sample was polished using a series of wheels covered by successively finer grit paper. The final polishing was done on a diamond wheel because the amount of aluminum, contained in the felt wheel, caused interference during preliminary analyses. In addition, polishing on a diamond wheel produces the smooth surface required for this kind of analysis.

The smooth surface is essential for microprobe analysis because unless the geometry of beam, sample surface and area of x-ray generation is the same from one area to the next on the sample, the resulting readings cannot be considered strictly comparable. If there is a pit on the surface very close to the electron beam/sample contact area, the subsurface bulb of x-ray generation will be closer to the surface on the side with the pit. More x-rays will escape from that area because fewer will be absorbed by the sample. If one of the diffracting crystals is in the path of increased radiation, it may transmit relatively more of its x-ray photons to the detector even though a greater amount of that element may not be present. The number of characteristic x-rays striking a crystal can be considered proportional to the concentration of the element in the sample only when the x-rays from that element are radiating roughly equally in all directions. This is no longer true when there is a pit in the surface. Söremark and Grøn (1966) state that surface projections may in addition cause irregular absorption of the x-rays.

After polishing, the sample receives a thin coat of carbon by vacuum evaporation. The carbon prevents a charge buildup on the surface of non-conductive samples. With a buildup of a negative charge on the surface the electron beam is repelled resulting in a large amount of electron backscatter rather than excitation of sample elements.

The samples were composed of differing proportions of cortical and cancellous bone. The rib sample was composed of both cortical and cancellous bone. The first radius cross section was taken from the proximal epiphysis and was composed of a very thin layer of cortical bone surrounding cancellous bone. The second radius cross section was taken just distal to the head of the radius and had a thicker layer of cortical bone surrounding an open medullary cavity containing some cancellous bone. The femoral cross section consisted entirely of cortical bone. The small number of cross sections was considered adequate since qualitatively different patterns between cortical and cancellous bone should be apparent with a sample of this size, if they actually exist.

Readings were taken across the bone samples. Figure 12 illustrates the cross section from the proximal epiphysis of the radius; the line indicates where the readings were made. Readings were taken every 100 microns along these lines. In later samples readings were taken every 300 microns since the distribution from such a sampling technique matched the more intense one. Two lines were scanned on both radius specimens while only one line was scanned on the femur specimen. Movement under the probe was along the Y-axis only, except in the areas of the medullary cavity. If the area chosen by this method turned out to be in plastic (resin) instead of bone the sample was moved along the X-axis until bone was encountered. The reading was taken at that position and then the sample was translated back to its original position for the next move along the Y-axis.

Instrumental conditions for the analysis were as follows: accelerating potential, 15 kili-electron volts; electron sample current, 0.02 micro-amps; beam diameter, 15 microns-normal incidence; three reflecting crystals; Ca-K_α line, P-$K_{\beta_{III}}$ line and Sr-L_α line.

Results

A plot of the counts of calcium, phosphorus and strontium resulting from sampling along a typical line across the bone is shown in Figure 13. This particular plot is that of a line positioned similar to A in Figure 12. The magnitude of the first five points represents the number of counts resulting from the excitation of the three elements in cortical bone; between area five and area twenty-nine the readings were taken on tra-

APPENDIX A   67

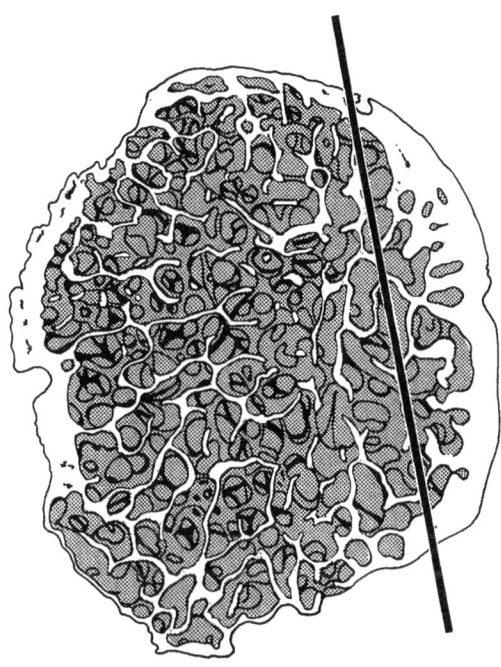

Fig. 12. Cross section of the head of a human radius. This section is typical of the sample used for the microprobe analysis. Readings were taken at 300 micron intervals along a line similar to the one shown on the right side of the figure.

becular bone in the marrow cavity and the last points are again counts taken on cortical bone. Although there is variation in the number of counts between sampling stations, there appears to be no observable difference in pattern between the cortical and trabecular areas.

If a large difference in mineral composition were present between cortical and trabecular bone, a higher coefficient of variation would be expected from those areas containing both cortical and trabecular bone when compared to those with only cortical bone. The coefficients of variation of the strontium, calcium and phosphorus readings from the areas of the different samples can be seen in Table 10. There is no consistent increase in the coefficient from areas of cortical bone to portions containing cortical plus trabecular bone. This indicates that a large difference in mineral composition does not exist between the two areas.

The elemental ratios were also plotted (see Fig. 14). Again, there appears to be no difference in pattern between cortical and cancellous bone. Aitken (1976) found that same lack of pattern differential in calcium/phosphorus ratios between cortical and trabecular bone, which

Table 10. Coefficients of variation of readings taken on lines across bone samples

	Rib			Radius				Femur
Sample Number	1	2	3	4	5	6	7	8
Type*	C	C+T	C+T	C	C+T	C	C+T	C
Strontium	15	15	11	14	16	12	10	11
Calcium	10	07	21	09	09	10	13	09
Phosphorus	10	08	11	18	19	09	15	29

C* = cortical
C+T = cortical + trabecular

supports the findings reported here.

The calcium and phosphorus values tend to covary rather closely with each other. This pattern was repeated in all the scans that were plotted. The strontium generally follows the pattern set by the other two elements but appears to vary much less than the calcium or the phosphorus. In areas of high or low mineralization it would be expected that the elemental concentrations of bone mineral would rise and fall together. On the other hand, strontium is incorporated into bone mineral as a 2^+ cation, the position usually filled by calcium. Therefore, both cations should increase with increased mineralization but not change identically since one replaces the other.

When a gram of bone is analyzed for the strontium and calcium concentrations, the ratio is fairly constant within one individual (Thurber et al. 1958). This consistency does not mean that every bone crystal within that individual contains a constant Sr/Ca ratio. The diameter of the electron beam is 15 microns and although the area of x-ray generation, as described in Appendix B, is larger than this, the areas being sampled are much closer to the size of a single crystal of bone mineral (100Å x 20-30Å, Neumann and Neumann 1969). The level of discrimination by the microprobe is fine enough that variation becomes apparent that is not observable on the macrolevel of the gram of bone.

Conclusions

There is variation observable in the amount of calcium, phosphorus and strontium when measured on the microlevel analyzed with the electron microprobe but no qualitative differences in this distribution appear between the cortical and cancellous areas of one individual bone. The coefficients of variation for the readings taken only on cortical bone are not lower in every case than the coefficients of variation for readings taken on combinations of cortical and cancellous bone. Thus, it is indicated that the differences within one bone are as small as

APPENDIX A

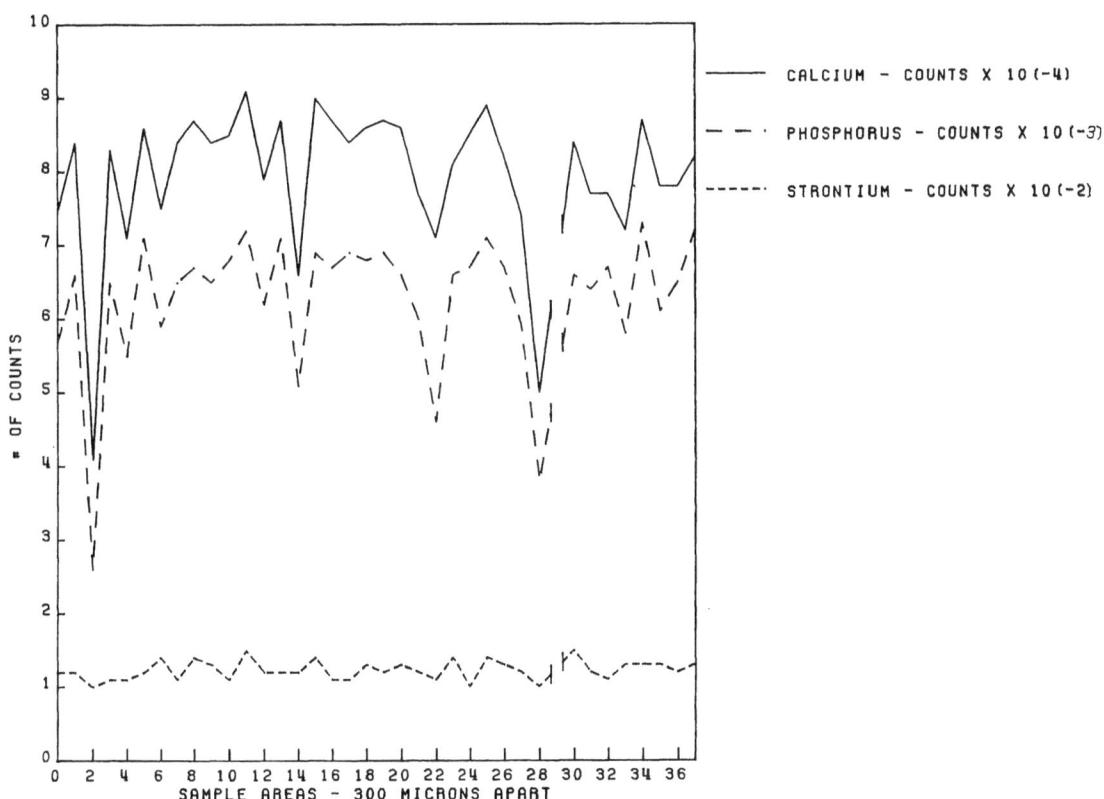

Fig 13. Digitized electrical output from a spectrophotometer. The output is proportional to concentration, therefore, a change in the number of counts, received from one element, indicates a change in the concentration of that element. The term sampling area refers to the area on which readings were recorded from the spectrophotometer. These areas were spaced 300 microns apart along a line similar to the one shown in Figure 12.

those between bones of one individual. Since the amount of bone analyzed by either atomic absorption or neutron activation analysis is much larger than that analyzed by microprobe the differences would be expected to average out producing one consistent value per bone whether cortical, trabecular or a combination of the two was used as the sample.

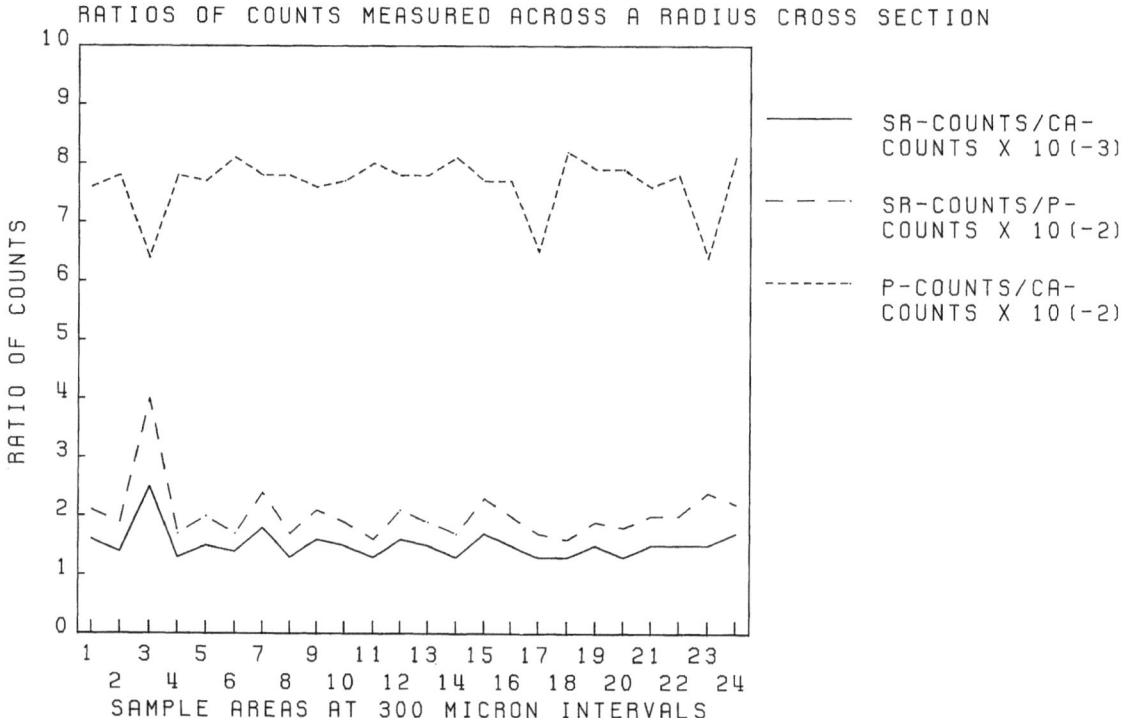

Fig. 14. Plot of the Sr/Ca, P/Ca, and Sr/P ratios of the digitized output from the spectrophotometer produced at each area analyzed by microprobe. Any change in a ratio between areas indicates a change in concentration in one or both elements. The sampling areas are the same as those referred to in Figure 13.

APPENDIX B

MICROPROBE ANALYSIS

In this method of analysis inner shell electrons of the elements composing the sample are excited by the energy transmitted to them from an electron beam. This energy knocks the inner electrons out of their orbits (ionization) creating a vacancy which is almost immediately filled by one of the outer shell electrons. As the outer shell electron moves down into the lower energy orbit, its excess energy is given off as an x-ray photon. The energy of this x-ray photon is characteristic of the element from which it originated. It is this relationship which provides the basis for microprobe analysis.

The energy of the x-ray photon is dependent upon the difference in energy between the vacated orbit and the orbit supplying the electron that fills the vacancy. For this reason several different x-rays may be emitted, all of which are characteristic of the element. If an electron is knocked out of the innermost orbit, a K x-ray results. Similarly, L and M x-rays result from removal of electrons from L and M orbits. Since a number of outer shell electrons can fill the vacancy, an alpha or beta sign with a number subscript following the shell letter is used to indicate the transition which took place. An element, excited by the electron beam will emit x-ray photons of several different energy levels, corresponding to the K_α, K_β, etc. transition which took place. The relative intensities of these emissions are characteristic of the particular element since the unique energy shell configuration of each element determines which shell transitions will be more likely.

Not all of the incident electrons dissipate their energy by ionizing electron shells. They may undergo several additional inelastic collisions before reduction to thermal velocity (heat energy). These additional collisions produce x-ray photons which are not characteristic of any particular element. This so-called background radiation, bremsstrahlung or white radiation, is proportional to the matrix atomic number and appears in the spectrum underlying the peaks produced by the characteristic x-ray emission. The separation of the energy produced by the element of interest from the background radiation may be difficult in the analysis of trace elements since the peak to background ratio may be low.

The kind of detection system, therefore, is an important consideration in trace element analysis. X-rays may be detected by an energy dispersive system or a wavelength dispersive system. In an energy dispersive system, the energy of each x-ray photon reaching the detector (lithium drifted silicon crystal) from the sample is identified. This information and the number of each type of photon striking the detector is stored in a multi-channel analyzer and is displayed as a spectrum composed of peaks. The height of the peak is determined by the number of photons striking the detector and is therefore proportional to the concentration of the element in the sample. The position of the peak is determined by the energy of the photon and is therefore indicative of the element producing it. In this type of system, however, "the pulses from the detector are small and require much more amplification, filtering to reduce noise and integration over a much longer pulse height..." (John Russ ms.). The spectrum is very useful for quick scanning to determine which elements are present in the sample but the wavelength dispersive system is more sensitive for quantitative measurement.

The wavelength dispersive method makes use of the wave like property of the x-ray. The wavelength of the x-ray photon is inversely proportional to its energy and is therefore characteristic of a particular element. A diffracting crystal is oriented in a particular position relative to the x-rays emitted from the sample. The crystal separates the x-ray photons of the element of interest from the rest since it is positioned so that only the photons of the element of interest are in phase when they are reflected from the crystal. This is important since if they are in phase, they are additive, producing an order of magnitude discernible by the detector.

The incoming x-rays strike different planes within the crystal lattice. The distances (d) between these planes are equal and are characteristic of the crystal. The characteristic x-rays which strike planes beneath the crystal surface will travel a greater distance (2a in Fig. 15) than those which reflect from the surface. $\sin \theta = a/d$, therefore, $2a = 2d\sin\theta$. The quantity, 2a, must equal an integer number of wavelengths (λ) if the reflected x-rays are to be in phase. It follows, then, that $n\lambda = 2d\sin\theta$. When the crystal is rotated so that it satisfies this Bragg condition, only the characteristic x-rays are reflected in phase to be detected and counted. Background interference is reduced when using the wavelength dispersive method, thus, sensitivity is increased. The three crystals used in this project were: ADP for calcium, ADP for strontium and RAP for phosphorus.

APPENDIX B

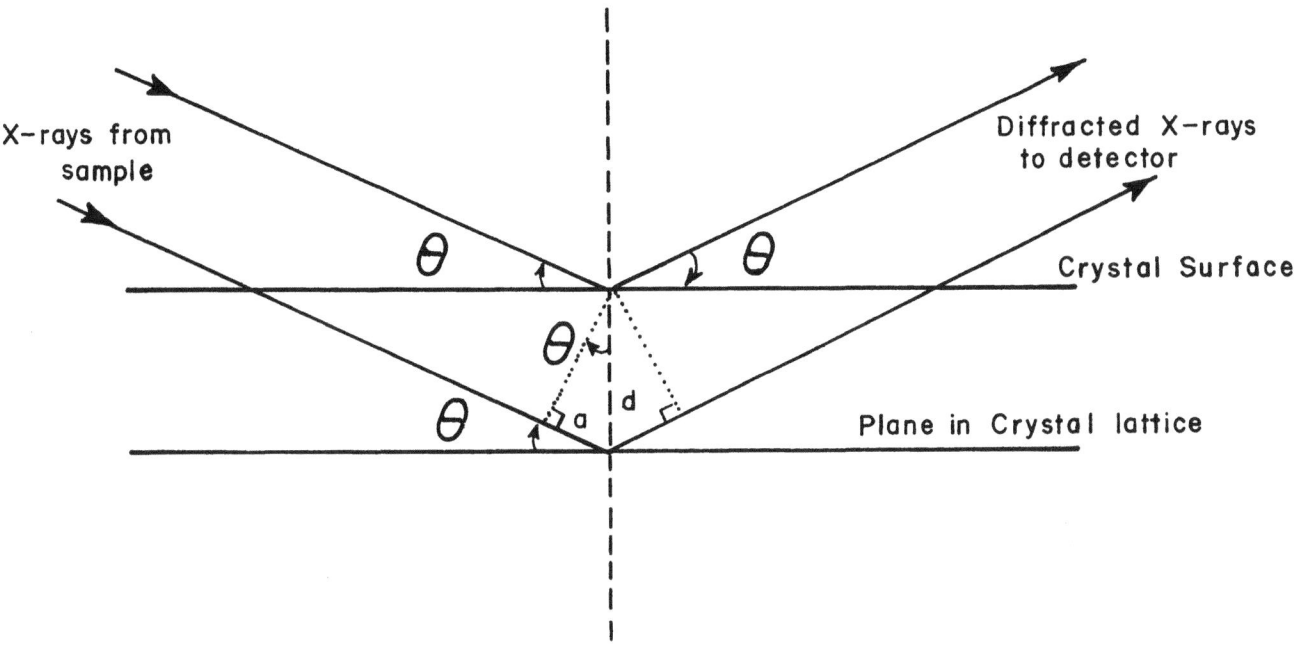

Fig. 15. According to Bragg's Law, incoming x-rays strike and are reflected from the crystal at an angle θ. Those which strike interior planes must travel an extra distance, $2a$. Since $\sin\theta = a/d$, where d is the distance between the crystal lattice planes, $2a = 2d\sin\theta$. For the reflected x-rays to be in phase, $2a$ must equal $n\lambda$ or $n\lambda = 2d\sin\theta$.

APPENDIX C

ATOMIC ABSORPTION SPECTOMETRY

Atomic absorption spectrometry (AAS), as the name implies, involves the measurement of the amount of energy (near ultra-violet or visible light) absorbed by particular atoms in a sample. Each element in the atomic state has a characteristic amount of energy which when absorbed will excite outer electrons out of their stable ground state without causing ionization. In AAS this characteristic energy is supplied by a hollow cathode lamp. Because both absorption and emission occur at the same characteristic wavelength for isolated atoms, the cathode is lined with the element of interest. When energized, the hollow cathode tube emits an energy characteristic of the element of interest (see Fig. 16). The atoms of this element within the sample absorb this energy in proportion to their concentration in that sample.

The sample, in liquid form, is aspirated through a small capillary tube into a flame. Because the components of the sample are in atomic form the atoms will absorb a portion of the energy supplied by the hollow cathode tube whose beam is directed down the line of the flame. The energy (light) not absorbed in the flame by the sample is directed, by means of a collimator, through a slit into a detector. The detector is situated to receive and count the intensity (I_t) of the incoming beam. The intensity (I_t) is the amount of the original light (I_o) which was not absorbed by the sample; so the absorbance (A) could be calculated:

$$A = \log \frac{I_o}{I_t}$$

This is not usually done because many things affect the linear relationship. Instead, a series of standards: a blank containing none of the element of interest and three or four other solutions with increasingly greater concentrations within the known linear range (given in the AAS manuals) are aspirated and the amount of deflection by the meter is noted. This calibration curve, checked often during a run, is used to calculate the concentration in an unknown sample. This works only if the curve is linear. The linear portion for strontium is between 0.0 and 7.0 microgram per milliter of solution (microgram equals 10^{-6} gram). Some of the AAS manuals listed 0.0-5.0 microgram per milliter as the linear range but my calibration curves were linear in the range stated previously. The concentration in an unknown is calculated by the following equation:

$$\frac{\text{Meter reading of unknown sample}}{\text{Meter reading of 1.0 } \mu g \text{ Sr/ml}} = \mu g \text{ Sr/ml unknown solution}$$

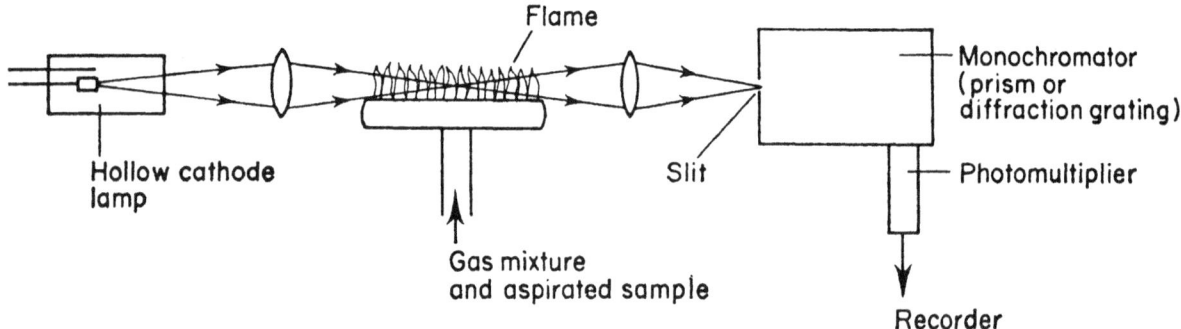

Fig. 16. The sample is aspirated through a capillary tube into the flame. The atoms of the elemnt of interest in the sample absorb some of the energy produced by the hollow cathode lamp while the remainder of the energy is directed to the detector. The difference between the amount of energy reaching the detector and the amount of energy produced by the hollow cathode lamp is proportional to the concentration of the element of interest within the sample. Figure Reprinted from: M.S. Tite, <u>Methods of Physical Examination in Archaeology</u>, Seminar Press (London and New York), 1972. Copyright by Academic Press Inc. (London) Ltd.

There are systematic errors in atomic absorption spectrometry due to the interaction between sample matrix and the excitation source (i.e., emission from the hollow cathode ray tube). Efficient absorption depends upon complete atomization of the sample. In the flame, stable compounds are formed between the strontium cation and phosphate or sulphate anions. Once this occurs, the outer electrons of strontium are no longer free to absorb the energy emitted from the hollow cathode tube. Absorption is lowered due to these interferences. Helsby (1974) has shown that the percent recovery of a known amount of strontium may be as low as 20% in an air-acetylene flame when calcium and phosphate are in the sample matrix. When Helsby used a nitrous-oxide acetylene flame (a much hotter flame) the percent recovery was around 40%. The atomic absorption manuals recommend the addition of a 1.0% solution of lanthanum because lanthanum will bond more readily than strontium with

APPENDIX C

the phosphate, thus removing the interference due to the strontium phosphate compound. Helsby reports that the percent recovery was raised from 20% to 50% when lanthanum was added. The remaining interference, Helsby attributes to the presence of calcium in the sample. In addition, at the levels found in bone ash, phosphate may still be a depressant on the level of abosrption, since Helsby reports a plateau of interference elimination at a concentration of lanthanum of 1.5%. In nitrous oxide-acetylene flames, the overall interference is less because the hotter flames breaks some of the molecular bonds, but the removal of phosphate as an interferent does not completely eliminate interferences because of the presence of caucium in bone.

If some of the interferences are lowered by using a hotter NO_2 flame, the problems due to ionization are increased. If the heat of the flame supplies enough energy to ionize the atoms in the sample, they will not absorb energy from the cathode ray tube in proportion to their concentration. Addition of a potassium salt (usually KCl) minimizes this problem since pottassium is more easily ionizable and removes some of this excess energy.

There are important advantages to the use of atomic absorption spectrometry. The equipment is relatively inexpensive so that several departments at many colleges and universities have atomic absorption set ups. The cost of a lamp, the reagents, glassware, etc. is probably all that would be required for beginning a project of trace element analysis. In addition, samples can be run fairly quickly. The time spent in sample preparation is about equal to that required in any method, but the time for actual analysis is much shorter than almost any other method. For these reasons, after a thorough consideration of the possibilities for errors, atomic absorption was used as one of the analytical techniques.

APPENDIX D

NEUTRON ACTIVATION ANALYSIS

Introduction

In this technique, slow neutrons are used to activate (energize) the atomic nuclei of the elements in the sample. If a neutron is added to an atom in the sample the formerly stable isotope is transformed into an unstable one (see Fig. 17). This unstable isotope emits a gamma ray as its means of returning to a stable state (as a new element). These gamma rays are characteristic of the original stable isotope, therefore, the number of gamma rays of a characteristic energy is proportional to the concentration of the original stable isotope in the sample.

There are several determinants of the probability of formation of a particular unstable isotope. One of these is the percent occurrence in nature of the parent stable isotope. Strontium has four such stable isotopes occurring in nature: Sr^{84} which comprises 0.56% of all stable strontium; Sr^{86} at 9.86%; Sr^{87} at 7.02% and Sr^{88} at 82.56%. The formation of the unstable isotope Sr^{85}, for example, depends partially on the natural occurrence of Sr^{84}. For this reason, the calculation of the concentration of total strontium in a sample requires consideration of both the amount of the unstable isotope and also the frequency of formation of that isotope. In addition, consideration must be given to the possibilities of the formation of other radioisotopes when activation occurs, the specific activity of the isotope and the length of the half-lives. Longer half-lives (not extreme) are sometimes preferable since interference from shorter-lived isotopes is minimized during gamma ray analysis. Sr^{87m} with a half-life equal to 2.8 hours is the isotope commonly used for Sr analysis by neutron activation analysis (NAA). Because its gamma rays must be counted quickly (30 minutes to two hours) there is quite a bit of interference from other isotopes with short half-lives (e.g., Na, Cl, Al, Mg). This is true especially in a complicated matrix like bone which has a large number of different constituents and where the element of interest is in such trace amounts. Such interference may account for the low replicability of previous NAA results for strontium in bone.

The isotope used for this analysis was Sr^{85}. Although the isotopic abundance of the parent isotope is not great, the specific activity is high (making counting easier and more reliable), and its relatively long half-life (65.2 days) eliminates interference from many shorter lived isotopes during energy separation. A long term irradiation (30 hours)

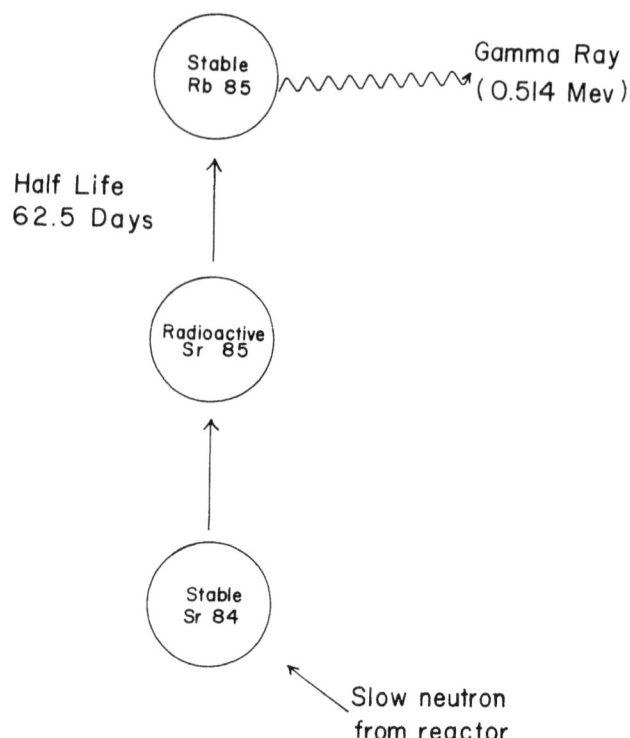

Fig. 17. A slow neutron from the reactor strikes the nucleus of a stable isotope Sr^{84}. This additional neutron transforms Sr^{84} into radioactive Sr^{85}. Within 62.5 days half of the Sr^{85}, produced in this manner, will emit a gamma ray with energy equal to 0.514 Kev. This emission transforms radioactive Sr^{85} into stable Rb^{85}. The number of gamma rays with energies of 0.514 Kev. is proportional to the number of Sr^{85}. In turn, the number of Sr^{85} is proportional to the original concentration of Sr^{84} in the sample.

is required, however, because the low abundance of Sr^{84} (0.56%) requires that a relatively large amount of it must be activated.

Irradiation

The neutron activation analysis was performed at the Phoenix Laboratories of the University of Michigan. The supply of neutrons (the neutron flux) is provided by a nuclear reactor. Exposure (irradiation) of the sample to the neutron flux may take place by sinking the samples into a position in the reactor's cooling pool for long term irradiations or by passing them briefly near the reactor core in a pneumatic tube system. Because long term irradiation was required, the bone samples were placed in the reactor pool.

Certain conditions had to be strictly controlled to avoid random

APPENDIX D 81

error intrusion. The amount of radiation had to be the same both among all samples and throughout all areas of one sample to insure that the concentration of the activated isotope was representative of the sample. This was accomplished by placing the sample in an area of the pool known to have a stable flux. In addition, the large container which houses the individual sample containers (sealed quartz tubing) was continually rotated in order to avoid any flux variation in the horizontal direction. The flux variation in the vertical direction has been found to be 15% per inch. Since the samples are 0.50 milligrams each, they are only about one quarter of an inch in height and are not likely to be affected by variation in the vertical direction.

Counting

The samples were cooled (i.e., shorter lived isotopes emitted their gamma rays which lowers the overall radioactivity of the sample) for at least three weeks by which time the activity of the Sr^{85} isotope is usually high enough for counting. The samples are then "counted"; that is, the number of gamma ray photons at each energy level is counted. The detector is a semiconductor counter; a germanium crystal which has lithium drifted through the crystal lattice (GeLi detector). When the gamma rays of all energy levels from the sample strike and are absorbed by the crystal, their energy is converted into an electrical pulse. The energy of the electrical pulse is characteristic of the isotope which originally produced it and the intensity (the number of pulses) is proportional to concentration. The GeLi detector transfers the various electrical impulses to a specialized computer, a multichannel analyzer (MCA), which keeps track of the number of pulses of each energy level detected by the GeLi crystal. The MCA is also called a pulse height analyzer because it digitizes a complete spectrum of all the energy levels produced by the sample. The energy levels displayed on a screen (in kili-electron volts which equal 10^3 electron volts) are along the x-axis and the height along the y-axis is the pulse height, proportional to concentration of that particular energy level. The energy resolution of this kind of detector is 15 times that of the type used less than ten years ago (thallium activated sodium iodide scintillation detector). It is this precision of energy resolution which makes strontium analysis possible now by neutron activation analysis without a chemical separation. Over twenty years ago, Harrison and Raymond (1955) advised the use of NAA for the assay of strontium in biological materials as a way to avoid the interference from calcium in the other commonly used analytical methods (especially atomic absorption spectrometry). At that time, however, they had to perform a chemical separation on all samples because the energy resolution was not fine enough. This step is no longer necessary.

During counting, certain precautions were taken. The samples were rotated continually and they were always situated in the same position relative to the detector so that errors due to geometry did not enter the calculations. Each sample was counted for two hours to make sure that enough gamma rays were emitted to represent the elemental composition of the sample. Because of the time delay between the first and last sample, the final calculation of concentration included a weight factor to offset the different counting times.

An Interference

The energy level used for determining the amount of strontium is at 514 kili-electron volts. The energy resolution for the GeLi detector is 2.3 kili-electron volts. At 511 kili-electron volts there is another peak. It is generally smaller than that produced by strontium and so appears as a shoulder on the 514 kili-electron volts peak. Two methods were used to separate the two peaks (a doublet). The first method was Gaussian doublet resolution which is a computer program that fit normal curves to the two peaks. The second method was an interference subtraction method which was worked out for this project by John Jones and Ward Rigot both of whom are members of the staff at the Phoenix Laboratory.

The interference subtraction method works as follows. There are several radionuclides known to emit a gamma ray with an energy of 511 kili-electron volts that can be produced in the neutron flux of a nuclear reactor. They are:

Isotope	Half-live
Cu^{64}	12.8 hours
Br^{80}	17.6 minutes
Zr^{89}	78.0 hours
Ni^{57}	36.0 hours
Co^{58}	71.3 days
As^{74}	17.5 days
Rh^{102m}	210.0 days
Na^{22}	2.6 years
Zn^{65}	245.0 days

Four of these elements have half-lives sufficiently shorter than strontium that their contribution can be considered to be minute (Cu, Br, Zr, and Mi). Cobalt occurs in bone at only 0.87 ppm (Bowen 1976) and arsenic at about 0.30 ppm (Bowen 1976). Both of these are in concentrations too low to be contributors to the doublet. Rhodium does not seem to have even been found in bone. Sodium does occur in fairly high levels in bone but the sodium 22 radioisotope occurs in too low a frequency to be a contributor. A scan of the complete energy spectrum (isotopes may emit at several different energy levels and an element may have more than one radioactive isotope) revealed that Zn^{65} was the most likely contributor.

The amount of zinc in the sample could be calculated using another energy peak and the known ratio of the two zinc lines would allow calculation of the amount of the 511/514 kili-electron volts doublet which was due to zinc. The other energy line of zinc (1115 kili-electron volts)

APPENDIX D

is, however, also a doublet, this time with scandium (1120 kili-electron volts). The amount of scandium in the sample was calculated using the 889.4 kili-electron volts scandium line and the known ratio between the two energy emission lines was used to determine the amount of the 1115/1120 doublet attributable to scandium. The calculated zinc intensity was then used to determine the amount of the 511/514 doublet attributable to zinc. Once this was subtracted out only the intensity of the strontium 514 kili-electron volt peak remained. In outline form:

a. Known ratio of intensities of $\dfrac{Sc(1120\ Kev.)}{Sc(889.4\ Kev.)}$ x intensity of Sc at 889.4 Kev. = intensity of Sc at 1120 Kev.

b. Intensity of doublet (Sc1120/Zn1115 Kev.) − intensity of Sc (from a above) = intensity of Zn at 1115 Kev.

c. Known ratio of intensities of $\dfrac{Zn(511\ Kev.)}{Zn(1115\ Kev.)}$ x intensity of Zn at 1115 Kev. (from b above) = intensity of Zn at 511 Kev.

d. Intensity of doublet (Zn 511/Sr514 Kev.) − intensity of Zn (from c above) = intensity of Xr at 514 Kev.

The number of emissions producing the curve, minus those in the shoulder, are compared with the standards and from this the parts strontium per million parts of bone ash is calculated:

$$\frac{PPM\ of\ the\ standard}{\frac{counts\ of\ std}{weight\ of\ std} \times \frac{weight\ of\ unknown}{counts\ of\ unknown}} = \text{parts per million of the unknown}$$

Chemical Separation

Method

In order to verify the results of either the Gaussian fit or of the interference subtraction method, a chemical separation was performed. The strontium was separated from the rest of the sample and recounted to make sure that only strontium emissions were being included in the parts per million calculation. The separation technique, described below, was suggested by Dr. K. Rengan of Eastern Michigan University.

The bone ash was removed from its quartz tubing sample containers and dissolved in one milliliter of concentrated nitric acid. To the dissolved bone ash 0.25 milliliter of a solution with a concentration of 20 milligrams strontium per milliliter of solution of $SrCO_3$ and 0.10 milliliter of a solution of zinc acetate were added as carriers to insure a complete reaction. To this solution five milliliters of 1 molar Na_2CO_3

were added. This was centrifuged and the supernatant was poured off. The precipitate was dissolved in a minimum amount of concentrated HNO_3. After cooling this solution in an ice bath, fuming HNO_3 was added until a precipitate formed. The solution was centrifuged and the supernatant poured off. The precipitate was washed once in fuming HNO_3, centrifuged and decanted. This last step should leave only the strontium since strontium is soluble in concentrated nitric acid but not in fuming nitric acid while the interferents are soluble in both. After the last wash, the precipitate was dissolved in deionized water and transferred to a test tube for counting, under the same conditions as before when it was in the solid bone ash form. The geometry of the sample to the detector was controlled and there was continuous rotation of the sample during counting. The removal of interferents was confirmed by the absence of a doublet at the 511/514 kili-electron volts energy portion of the spectrum. Only the 514 kili-electron volt peak was present.

Determination of Recovery Success

After counting, the recovery fraction resulting from the above procedure was determined by measuring the amount of strontium carrier that remained after the separation. An insignificant amount of the original stable strontium in the bone ash sample should remain because of the long irradiation time (30 hours); therefore, all the stable strontium present in the sample should be due to the carrier. An aliquot, 0.25 milliliter of the liquid sample was taken from each test tube and placed in a polypropylene sample container. The caps of these containers were heat sealed to prevent leakage. The samples were arranged around the outside of a "rabbit" sample holder used in the pneumatic tube system for short-term irradiations. The samples were irradiated six times for ten seconds each time. The "rabbit" sample holder was rotated end for end between each irradiation. Prior experimentation had demonstrated that this technique kept the variation in counts due to flux differential down to $\bar{X} \pm 0.03$. The Sr^{85} isotope could not be used for this counting because too much of it still remained from the original irradiation of the bone ash. Instead, Sr^{87m} was used. The large amount of strontium in the sample (added as carrier) minimizes the problems mentioned above concerning the use of Sr^{87m}, involved in using this shorter-lived isotope. Each sample was counted for 120 seconds. The ratio of the amount of strontium added as carrier to the amount counted after this short-term irradiation is the fraction recovered by chemical separation. Adjustments were made for the difference in some samples of the time lag between irradiation and counting.

Parts Per Million Calculation

This recovery fraction was used in weighting the emission counts from Sr^{85} isotope after the chemical separation. After proper weighting, the parts per million strontium were calculated by comparison with a standard as outlined above. The parts per million values for strontium calculated by the Gaussian peak fitting program, the interference sub-

APPENDIX D

traction and chemical separation methods are listed in Table 11. The results of the chemical separation are probably the closest to the "true" value of strontium reached by neutron activation analysis, except for sample #37745 where the percent recovery is too high to be relied upon. For this reason the values of the Gaussian Doublet Resolution and the interference subtraction methods are shown as percentages of the chemical separation values. The interference subtraction method produces values which are much closer to the best approximation of the true value than does the Gaussian curve fitting program; therefore the interference subtraction method was used to separate the Zn 511 kili-electron volt peak from the Sr 514 kili-electron volt peak.

Advantages and Disadvantages

The advantage to the use of neutron activation is the lack of matrix interference, so much a problem in atomic absorption spectrometry. After comparison with the results of the chemical separation it would appear that instrumental neutron activation analysis is a very accurate method of assaying bone strontium. Without a nationally recognized standard for bone there is no way to check this but is seems a reasonable conclusion. The disadvantages of NAA include the cost and the turnover time for samples. The time from submission of a set of samples to the time results are available is a minimum of three weeks and more than twice that long is not unusual because of the scheduling needs for computer time which is required for both counting and for data reduction. Therefore, although I believe the results are more accurate and that a subset of the total sample should always be analyzed by NAA, the bulk of the samples should be analyzed by atomic absorption spectrometry.

Table 11. Comparison of results produced by different data reduction methods. All samples analyzed by NAA

Sample #	Gaussian	Interference Subtraction	Chemical Separation	(Percent Recovery)	Gaussian/ Chem. Sep.	Int. Sub./ Chem. Sep.
CH-T37 #4	1103.07	823.70	846.3	(74.8%)	130.0%	97.3%
CH-T24 #50	1491.17	1255.09	1276.8	(87.8%)	116.8%	98.3%
Oaxaca #2	855.83	739.73	742.5	(76.9%)	115.3%	99.6%
CH-T37 #5	806.17	643.90	536.9	(97.6%)	150.2%	119.9%
CH-T25 #8	859.70	700.52	718.9	(77.9%)	119.6%	97.4%
CH-T24 #6	882.71	778.84	753.6	(69.8%)	117.1%	103.3%
CH-T4E #1	796.07	736.90	658.2	(82.5%)	120.9%	112.0%
CH-T37 #1	1367.21	1185.20	1079.2	(80.8%)	126.7%	109.8%
CH-T25 #50	934.46	642.49	681.8	(88.0%)	137.1%	94.2%
CH-T25 #2	729.26	522.25	591.9	(89.2%)	123.2%	88.2%

BIBLIOGRAPHY

Aitken, J.M.
  1976 Factors Affecting the Distribution of Zinc in the Human Skeleton. Calcified Tissue Research 20(1):23-30.

Alexander, G.V., and R.E. Nusbaum
  1959 The Relative Retention of Strontium and Calcium in Human Bone Tissue. Journal of Biological Chemistry 234:418-21.

Alexander, G.V., R.E. Nusbaum, and N.S. MacDonald
  1956 The Relative Retention of Strontium and Calcium in Bone Tissue. Journal of Biological Chemistry 218:911-19.

Bang, S., and C.A. Baud
  1972 Topographic Distribution of Strontium and its Incorporation into Bone Mineral Substance in Vivo. Proceedings of the International Conference on X-Ray Optics and Microanalysis, Sixth annual, G. Shinoda, ed., pp. 841-45. University of Tokyo Press, Tokyo, Japan.

Barnard, A.J. Jr., and R.W. Dudley
  1973 Tracing the Elements. Industrial Research 15(3):34-7.

Bass, W.M., D.R. Evans, R.L. Jantz, and D.H. Ubelaker
  1971 The Leavenworth Site Cemetery: Archaeology and Physical Anthropology. University of Kansas Publications in Anthropology 2.

Bedford, J., G.E. Harrison, W.H.A. Raymond, and A. Sutton
  1960 The Metabolism of Strontium in Children. British Medical Journal 1:589-92.

Berg, A.
  1972 Metabolism of Calcium and Strontium in Fresh Water Fish. IV. Calcium and Strontium Relations in Fishes of Two Northern Italian Lakes and Their General Radioecological Implications. Memorie dell'Instituto Italiano di Idrobiologia Dott. Marco de Marchi 29:145-67.

Binford, L.R.
  1972 An Archaeological Perspective. New York: Seminar Press.

Binford, S.R.
  1968 Early Upper Pleistocene Adaptations in the Levant. American Anthropologist 70:707-17.

Boaz, N.T., and J. Hampel
1978 Strontium Content of Fossil Tooth Enamel and Diet in Early Hominids Journal of Paleontology 52(4):928-33.

Bowen, H.J.M.
1976 Trace Elements in Biochemistry. London: Academic Press.

Bowen, H.J.M., and J.A. Dymond
1955 Strontium and Barium in Soils and Plants. Proceedings of the Royal Society (London) Series B 144:355-68.

Boyde, A., V.R. Switzur, and R.W. Fearnhead
1961 Application of the Scanning Electron-Probe X-Ray Microanalyzer to Dental Tissues. Journal of Ultrastructure Research 5:201-07.

Brace, C.L.
1973 Environment, Tooth Form and Size in the Pleistocene. In: Man in Evolutionary Perspective, edited by C.L. Brace and J. Metress, pp. 315-23. New York: John Wiley and Sons Inc.

1977 Biological Parameters and Pleistocene Hominid Life-Ways. Paper presented at the Burg Wartenstein Symposium No. 75.

Brown, A.B.
1973 Bone Strontium as a Dietary Indicator in Human Skeletal Populations. Ph.D. Dissertation, University of Michigan, Ann Arbor.

1974 Bone Strontium as a Dietary Indicator in Human Skeletal Populations. University of Wyoming Contributions to Geology 13(2): 47-8.

Bryant, F.J. and J.F. Loutit
1961 Human Bone Metabolism Deduced From Strontium Assays. Atomic Energy Research Establishment-R3718, London:H.M.S.O.

Buikstra, J.E.
1976 The Caribou Eskimo: General and Specific Disease. American Journal of Physical Anthropology 45(3):351-68.

1977 Biocultural Dimensions of Archaeological Study: A Regional Perspective. In: Biocultural Adaptation in Prehistoric North America, edited by R.L. Blakely. Southern Anthropological Society Preceedings 11:67-84.

Childe, V.G.
1951 Social Evolution. London: Watts & Co.

Christenson, A.L., and D.W. Read
1977 Numerical Taxonomy, R-Mode Factor Analysis and Archaeological Classification. American Antiquity 42(2):163-79.

BIBLIOGRAPHY

Comar, C.L., R.S. Russell, and R.H. Wasserman
   1957 Strontium-Calcium Movement From Soil to Man. Science 126;485-92.

Comar, C.L., and R.H. Wasserman
   1964 Strontium. In: Mineral Metabolism, Vol 2, Part A, pp. 523-72, edited by C.L. Comar and F. Bronner.

Comar, C.L., I.B. Whitney, and F.W. Lengemann
   1955 Comparative Utilization of Dietary Sr^{90} and Calcium by Developing Rat Fetus and Growing Rat. Proceedings of the Society for Experimental Biology and Medicine 88:232-36.

Cormack, R.M.
   1971 A Review of Classification. Journal of the Royal Statistical Society, Series A 134;321-67.

Cyphers, A.
   1975 Preclassic Ceramic Chronology at Chalcatzingo, Morelos, Mexico. M.A. Thesis, University of Wisconsin, Milwaukee.

Dart, R.A.
   1955 Cultural Status of the South African Man-Apes. Annual Report of the Smithsonian Institution:317-38.

DeVore, I., and S.L. Washburn
   1963 Baboon Ecology and Human Evolution. In: African Ecology and Human Evolution, edited by F.C. Howell and F. Bourliere, pp. 335-67. Chicago: Aldine.

Fisher, G.L., L.G. Davies, and L.S. Rosenblatt
   1976 The Effects of Container Composition, Storage Duration, and Temperature on Serum Mineral Levels. National Bureau of Standards Special Publication 422:575-92.

Fleischer, M.
   1953 Recent Estimates of the Abundance of the Elements in the Earth's Crust. U.S. Geological Survey Circular 285.

Frank, R.M., M. Capitant, and J. Goni
   1966 Electron Probe Studies of Human Enamel. Journal of Dental Research Supplement 45(3):672-82.

Garn, S.M.
   1966 Malnutrition and Skeletal Development in the Preschool Child. Pre-School Child Malnutrition, National Academy of Sciences-Council Washington, D.C.

Gilbert, R.I.
   1975 Trace Element Analyses of Three Skeletal Amerindian Populations at Dickson Mounds. Ph.D. Dissertation, University of Massachusetts, Amherst.

Gorsuch, T.T.
  1976 Dissolution of Organic Materials. National Bureau of Standards Special Publication 422:491-508.

Grennes-Ravitz, R.A., and G.H. Coleman
  1976 The Quintessential Role of Olmec in the Central Highlands of Mexico: A Refutation. American Antiquity, 41(2):196-206.

Grove, D.C.
  1968 Chalcatzingo, Morelos, Mexico: a Reappraisal of the Olmec Rock Carvings. American Antiquity 33(4):486-91.

  1970 The Morelos Formative: Cultural Stratigraphy and Implications. Paper presented at the Society for American Archaeology Meetings, Mexico City.

  1973 Olmec and "Olmec": New Data from Chalcatzingo. Paper presented at the American Anthropological Association Meetings, Toronto.

  1974 Chalcatzingo, Is it Really Olmec? Paper presented at the American Anthropological Association Meetings, Mexico City.

  1975 Formative Period Interaction Spheres in Morelos and the Valley of Mexico. Paper presented at the American Anthropological Association Meetings, San Francisco.

Grove, D.C., K.G. Hirth, D.E. Buge, and A.M. Cyphers
  1976 Settlement and Cultural Development at Chalcatzingo. Science, 192:1203-10.

Hall, T.
  1968 Some Aspects of the Microprobe Analysis of Biological Specimens. In: Quantitative Electron Probe Microanalysis, edited by K.F.J. Heinrich. National Bureau of Standards Special Publication 298:269-99.

Hare, P.E.
  1976 Organic Geochemistry of Bone and Its Relation to the Survival of Bone in the Natural Environment. Paper presented at the Burg Wartenstein Symposium 69.

Harris, M.
  1968 The Rise of Anthropological Theory. New York: Thomas Y. Crowell Co.

Harrison, G.E., and W.H.A. Raymond
  1955 The Estimation of Trace Amounts of Barium or Strontium in Biological Material by Activation Analysis. Journal of Nuclear Energy 1:290-98.

Hatch, J.W.
1976 Status in Death: Principles of Ranking in Dallas Culture Mortuary Remains. Ph.D. Dissertation, Pennsylvania State University, University Park.

Hatch, J.W., and P.S. Willey
1974 Stature and Status in Dallas Society. Tennessee Archaeologist, 30(2):107-31.

Haviland, W.A.
1967 Stature at Tikal, Guatemala: Implications for Ancient Maya Demography and Social Organization. American Antiquity, 32(3):316-25.

Helsby, C.A.
1974 Determination of Strontium in Human Tooth Enamel by Atomic Absorption Spectrometry. Analytica Chimica Acta 69:259-65.

Hirth, K.G.
1977 Interregional Trade and the Formation of Prehistoric Gateway Communities. American Antiquity 43(1):35-45.

Hodges, R.M., N.S. MacDonald, R. Nusbaum, R. Stearns, F. Ezmirlian, P. Spain, and C. MacArthur
1950 Strontium Content of Human Bones. Journal of Biological Chemistry 185:519-24.

Ivanov, V.K., and V.I. Pashkova
1974 Establishment of Species Identification of Bone Fragments by Emission Spectrographic Analysis, Preliminary Report. Sudebno-Meditsinskaya Ekspertiza, Ministerstvo Zdravookhraneniya SSSR 17(3):13-14.

Jolly, C.J.
1970 The Seed-Eaters: A New Model of Hominid Differentiation Based on a Baboon Analogy. Man 5(1):5-26.

King, G.E.
1975 Socioterritorial Units Among Carnivores and Early Hominids. Journal of Anthropological Research 31(1):69-87.

1976 Society and Territory in Human Evolution. Journal of Human Evolution 5:323-32.

Kulebakina, L.G.
1975 Strontium-90 in the Cystoseiric Biocenosis of the Black Sea Shelf Zone. In: Self-Purification, Bioproductivity, and Protection of Reservoirs and Currents of Water in the Ukraine, edited by A.V. Topachevskii, pp. 102-4, Kiev, USSR:Naukova Dumka.

Lengeman, F.W.
　1963　Over-All Aspects of Calcium and Strontium Absorption. In: The Transfer of Calcium and Strontium Across Biological Membranes, edited by R.H. Wasserman. New York: Academic Press.

Leonard, F., and D.I. Scullin
　1969　New Mechanism for Calcification of Skeletal Tissue. Nature 224:1113-15.

Long, C. (ed.)
　1961　Biochemist's Handbook. Princeton, New Jersey: Van Nostrand Co.

Lough, S.A., J. Rivera, and C.L. Comar
　1963　Retention of Strontium, Calcium and Phosphorus in Human Infants. Proceedings of the Society for Experimental Biology and Medicine, 112:631.

Loutit, J.F.
　1967　Strontium-90 From Fall-Out in Human Bone. In: Strontium Metabolism, edited by J.M.A. Lenihan, J.F. Loutit and J.G. Martin. pp. 41-5. New York: Academic Press.

Malcolm, L.A.
　1974　Ecological Factors Relating to Child Growth and Nutritional Status. In: Nutrition and Malnutrition: Identification and Measurement, edited by A.F. Roche and F. Falkner. pp. 329-52. New York: Plenum Press.

Marchall, J.H., J. Liniecki, E.L. Lloyd, G. Marotti, C.W. Mays, J. Rundo, H.A. Sissons, and W.S. Snyder
　1973　Alkaline Earth Metabolism in Adult Man. Health Physics 24:125-221.

Marquardt, W.H., and P.J. Watson
　1977　Excavation and Recovery of Biological Remains from Two Archaic Shell Middens in Western Kentucky. Bulletin of the Southeastern Conference.

Marx, E.
　1977　The Tribe as a Unit of Subsistence: Nomadic Pastoralism in the Middle East. American Anthropologist 79(2):343-63.

McLean, F.C., and M.R. Urist
　1968　Bone: An Introduction to the Physiology of Skeletal Tissue. 3rd ed. Chicago: University of Chicago Press.

Meehan, B.
　1977a　Man Does Not Live by Calories Alone: The Role of Shellfish in a Coastal Cuisine. In: Sunda and Sahul, Prehistoric Studies in S.E. Asia, Melanesia and Australia, edited by J. Allen, J. Golson, and R. Jones, pp. 493-531. London: Academic Press

BIBLIOGRAPHY

Meehan, B.
  1977b Hunters by the Seashore. Journal of Human Evolution 6(4):363-70.

Mellors, R.C., G.C. Kenneth, and T. Solberg
  1966 Quantitative Analysis of Calcium/Phosphorus Molar Ratios in Bone Tissue with the Electron Probe. In: Electron Microprobe, Proceedings of a Symposium, Washington, D.C. 1964. pp. 834-40.

Menzel, R.G., and W.R. Heald
  1959 Strontium and Calcium Contents of Crop Plants in Relation to Exchangeable Strontium and Calcium of the Soil. Soil Sciences Society of America Proceedings 23:110-12.

Merry, M.P.
  1975 Investigation of a Middle Formative Area of Burials: Chalcatzingo, Morelos, Mexico. M.A. Thesis, University of the Americas, Mexico City, Mexico.

Morrison, G.H.
  1976 Interpretation of Accuracy of Trace Element Results for Biological Materials. National Bureau of Standards Special Publication 422:65-78.

Murphy, T.J.
  1976 The Role of the Analytical Blank in Accurate Trace Analysis. National Bureau of Standards Special Publications 422:509-41.

Nelson, D.J.
  1967 Microchemical Constitutents in Contemporary and Pre-Columbian Clamshell. In: Quaternary Palaeoecology, edited by E.J. Cushing and H.E. Wright, Jr. pp. 185-204. New Haven: Yale University Press.

Neuman, W.F., R. Bjornerstedt, and B.J. Mulryan
  1963 Synthetic Hydroxyapatite Crystals. II. Aging and Strontium Incorporation. Archives of Biochemistry and Biophysics 101:215-24.

Neuman, W.G., and M.W. Neuman
  1969 The Chemical Dynamics of Bone Mineral. Chicago:University of Chicago Press.

Odum, E.P.
  1971 Fundamentals of Ecology. 3rd ed. Philadelphia:W.B. Saunders Co.

Odum, H.T.
  1951 The Stability of the World Strontium Cycle. Science 114:407-11.

  1957 Strontium in Natural Waters. Texas University Institute of Marine Science Publications 4(2):22-37.

Ophel, I.L.
　1963　The Fate of Radiostrontium in a Freshwater Community. In: Radioecology, edited by V. Schultz and A.W. Klement, pp. 213-16. London: Chapman and Hall.

Parker, R.B.
　1976　Trace Elements in Bones as Paleobiological Indicators. Paper presented at the Burg Wartenstein Symposium 69.

Parker, R.B., and H. Toots
　1970　Minor Elements in Fossil Bones. Geological Society of America Bulletin 81:925-32.

Peebles, C.S.
　1971　Moundville and Surrounding Sites: Some Structural Considerations of Mortuary Practices II. American Antiquity Memoir 25:68-91.

　1974　Moundville: The Organization of a Prehistoric Community and Culture. Ph.D. Dissertation, University of California, Santa Barbara.

Peebles, C.S., and S.M. Kus
　1977　Some Archaeological Correlates of Ranked Societies. American Antiquity 42(3):421-48.

Perkin-Elmer Corporation
　1971　Analytical Methods for Atomic Absorption Spectrophotometry. Norwalk, Connecticut.

Posner, A.S.
　1973　Bone Mineral on the Molecular Level. Federation Proceedings of the American Society for Experimental Biology 32(9):1933-37.

Reeve, J., and R. Hesp
　1976　A Model-Independent Comparison of the Rates of Uptake and Short Term Retention of Calcium-47 and Strontium-85 by the Skeleton Journal of Calcified Tissue Research 22:183-89.

Robbins, L.M.
　1977　The Story of Life Revealed by the Dead. In: Biocultural Adaptation in Prehistoric North America, edited by R.L. Blakely. Southern Anthropological Society Proceedings 11:10-26.

Robinson, J.T.
　1963　Adaptive Radiation in the Australopithecines and the Emergence of Man. In: African Ecology and Human Evolution, edited by F. Howell and F. Bourlière, pp. 385-416. Chicago: Aldine Publishers.

BIBLIOGRAPHY

Rosenthal, H.L.
  1963 Uptake, Turnover and Transfer of Bone Seeking Elements in Fishes. New York Academy of Science Annual 109:278-93.

Russ, J.C.
  1974 X-Ray Microanalysis in the Biological Sciences. Journal of Submicroscopy and Cytology 6:55-79.

  n.d. Principles and Instrumentation of Microanalysis in Scanning Electron Microscopy and Transmission Electron Microscopy.

Saxe, A.A.
  1970 Social Dimensions of Mortuary Practices. Ph.D. Dissertation, University of Michigan, Ann Arbor.

Schaller, G.B., and G.R. Lowther
  1969 The Relevance of Carnivore Behavior to the Study of Early Hominids. Southwestern Journal of Anthropology 25(4):307-41.

Schoeninger, M.J., and C.S. Peebles
  n.d. The Effects of Molluscs on the Use of Strontium as an Indicator of Prehistoric Human Diet. Unpublished manuscript in authors' possession, University of Michigan Museum of Anthropology, Ann Arbor.

Schroeder, H.A., I.H. Tipton, and A.P. Nason
  1972 Trace Metals in Man: Strontium and Barium. Journal of Chronic Diseases 24:491-517.

Service, E.R.
  1971 Primitive Social Organization. 2nd ed. New York: Random House.

Simons, E.
  1977 Ramapithecus. Scientific American 236(5):28-35.

Smith, B.D.
  1975 Middle Mississippi Exploitation of Animal Populations. Museum of Anthropology, University of Michigan, Anthropological Papers 57.

Smith, C.B., and D.A. Smith
  1976 An X-Ray Diffraction Investigation of Age Related Changes in the Crystal Structure of Bone Apatite. Journal of Calcified Tissue Research 22(2):219-26.

Sneath, P.H.A., and R.R. Sokal
  1973 Numerical Taxonomy. San Francisco: W.H. Freeman and Co.

Sokol, R.R., and F.J. Rohlf
  1969 Biometry. San Francisco: W.H. Freeman and Co.

Söremark, R., and P. Grøn
　1966　Chloride Distribution in Human Dental Enamel as Determined by Electron Probe Microanalysis. Archives of Oral Biology 2:861-66.

Sowden, E.M., and S.R. Stitch
　1957　Trace Elements in Human Tissue. Estimation of the Concentrations of Stable Strontium and Barium in Human Bone. Biochemical Journal 67:104-09.

Spadaro, J.A., R.O. Becker, and C.H. Bachman
　1970　The Distribution of Trace Metal Ions in Bone and Tendon. Journal of Calcified Tissue Research 6:49-54.

Speecke, A., J. Hoste, and J. Versieck
　1976　Sampling of Biological Materials. National Bureau of Standards Special Publication 422:299-310.

Spores, R.
　1965　The Zapotec and Mixtec at Spanish Contact. In: Handbook of Middle American Indians, 3(Pt. 2):962-87.

Szpunar, C.
　1977　Atomic Absorption Analysis of Archaeological Remains: Human Ribs from Woodland Mortuary Sites. Ph.D. Dissertation, Northwestern University, Evanston.

Teleki, G.
　1974　Chimpanzee Subsistence Technology: Materials and Skills. Journal of Human Evolution 3:575-94.

　1975　Primate Subsistence Patterns: Collector, Predator and Gatherer-Hunter. Journal of Human Evolution 4:125-84.

Termine, J.D., R.A. Peckauskas, and A.S. Posner
　1970　Calcium Phosphate Formation in Vitro II. Effects of Environment on Amorphous-Crystalline Transformation. Archives of Biochemistry and Biophysics 140:318-25.

Termine, J.D., and A.S. Posner
　1967　Amorphous/Crystalline Interrelationships in Bone Mineral. Journal of Calcified Tissue Research 1:8-23.

Thurber, D.L., J.L. Kulp, E. Hodges, P.W. Gast, and J.M. Wampler
　1958　Common Strontium Content of the Human Skeleton. Science 128:256-7.

Toots, H., and M.R. Voorhies
　1965　Strontium in Fossil Bones and the Reconstruction of Food Chains. Science 149:854-55.

Turekian, K.K., and J.L. Kulp
　1956　Geochemistry of Strontium. Geochemica Cosmochimica Acta 10:245.

BIBLIOGRAPHY

Vose, P.B., and H.V. Koontz
  1955 The Uptake of Strontium by Pasture Plants and its Possible Significance in Relation to the Fall-Out of Strontium-90. Nature 183:1447-8.

Wesch, H., and A. Bindl
  1976 Analysis of 11 Elements in Biological Material. Comparison of Neutron Activation Analysis and Atomic Absorption Analysis. National Bureau of Standards Special Publication 422:231-38.

Wessen, G., F. Ruddy, C. Gustafson, and H. Irwin
  1977 Bone Strontium and Barium as Indicators of Diet and Environment. Paper presented at the Annual Meeting of the Society for American Archaeology, New Orleans

Whallon, R., C.S. Peebles, and S.M. Kus
  1975 Experimenting with the Numerical Analysis of Shape in Pottery Vessels. Paper presented at the 41st InternationalCongress of Americanists, Mexico City.

White, L.A.
  1949 The Science of Culture. New York:Grove Press, Inc.

Wright, H.T.
  1977 Recent Research on the Origin of the State. Annual Review of Anthropology 6:379-97.

Wyckoff, R.W.G., and A.R. Doberenz
  1968 The Strontium Content of Fossil Teeth and Bones. Geochimica Cosmochimica Acta 32:109-15.

Yablonskii, M.F.
  1971 Use of Differences in Bone Mineral Content for Identification of Corpses. Sbornik Nauchngkh Trudor Vinnitskogo Gosudarstvennogo Meditsinskogo Instituta. Collection of Scientific Works of the Vinnitsa State Medical Institute 14:368-74.

  1973 Identificational Significance of Major and Trace Elements of Human Long Tubular Bones. Sudebno-Meditsinskaya Ekspertiza, Ministerstvo Zdravookhraneniya SSSR . Forensic-Medical Arbitration, Ministry of Public Health of the USSR 16:16-18.

www.ingramcontent.com/pod-product-compliance
Lightning Source LLC
Jackson TN
JSHW052242110426
100741JS00005B/26